83 Structure and Bonding

Editors:
M. J. Clarke, Chestnut Hill, MA
J. B. Goodenough, Austin, TX • C. K. Jørgensen, Genève
D. M. P. Mingos, London • J. B. Neilands, Berkeley, CA
G. A. Palmer, Houston, TX • P. J. Sadler, London
R. Weiss, Strasbourg • R. J. P. Williams, Oxford

Springer-Verlag Berlin Heidelberg GmbH

Iron-Sulfur Proteins
Perovskites

With contributions by
I. Bertini, S. Ciurli, C. Luchinat,
W. J. A. Maaskant

With 34 Figures and 5 Tables

 Springer

ISBN 978-3-662-14887-7

Library of Congress Cataloging-in-Publication Data
Bioorganic chemistry / with contributions by I. Bertini ... [et al.].
p. cm. -- (Structure and bonding ; 83) Previously published: Includes bibliographical
references and index.
ISBN 978-3-662-14887-7   ISBN 978-3-540-49188-0 (eBook)
DOI 10.1007/978-3-540-49188-0

1. Bioinorganic chemistry. I. Bertini, Ivano. II. Series.
QP531.B543 1995   574.19'214--dc20   95-12983 CIP

© Springer-Verlag Berlin Heidelberg 1995
Originally published by Springer-Verlag Berlin Heidelberg New York in 1995
Softcover reprint of the hardcover 1st edition 1995

Typesetting: Macmillan India Ltd., Bangalore-25, India
SPIN: 10101060   51/3020 - 5 4 3 2 1 0 - Printed on acid-free paper

Editorial Board

Attention all "Structure and Bonding" readers:

A file with the complete volume indexes Vols. 1 through 83 in delimited ASCII format is available for downloading at no charge from the Springer EARN mailbox. Delimited ASCII format can be imported into most databanks.

The file has been compressed using the popular shareware program "PKZIP" (Trademark of PK ware Inc., PKZIP is a available from most BBS and shareware distributors).

This file is distributed without any expressed or implied warranty.

To receive this file send an e-mail message to:
SVSERV@DHDSPRI6.BITNET.

The message must be: "GET /CHEMISTRY/SB_V1.ZIP".

SVSERV is an automatic data distribution system. It responds to your message. The following commands are available:

HELP	returns a detailed instruction set for the use of SVSERV,
DIR (*name*)	returns a list of files available in the directory "name",
INDEX (*name*)	same as "DIR",
CD <*name*>	changes to directory "name",
SEND <*filename*>	invokes a message with the file "filename",
GET <*filename*>	same as "SEND".

Table of Contents

The Electronic Structure of FeS Centers in Proteins and Models
A Contribution to the Understanding of Their Electron Transfer Properties

Ivano Bertini[1], Stefano Ciurli[2], and Claudio Luchinat[2]

[1] Department of Chemistry, University of Florence, Via Gino Capponi, 7, 50121 Florence (Italy)
[2] Institute of Agricultural Chemistry, University of Bologna, Viale Berti Pichat, 10, 40127 Bologna (Italy)

Iron-sulfur polymetallic centers pose interesting questions on the oxidation state of each metal ion, on the possibility of electron delocalization over the metal ions versus the possibility of existence of equilibria among different localized-valence states, and on the factors determining if, and to what extent, electron delocalization is present. By interpreting the hyperfine coupling of unpaired electrons with ^{57}Fe as well as with ^1H, and by analyzing the EPR spectra, the valence distributions of Fe–S polymetallic centers in proteins and model systems are obtained. The systems analyzed here are: $[Fe_2S_2]^{2+/+}$, $[Fe_3S_4]^{+/0}$, $[Fe_4S_4]^{3+/2+/+}$. The understanding of the parameters describing the exchange coupling within pairs of spin vectors, as well as the double exchange coupling, is attempted on the basis of the oxidation states of the various ions. Upon proper scaling of these parameters, we show that it is possible to transfer them from one polymetallic center to another. The electronic properties are tentatively related to the microscopic reduction potentials of each iron in the center.

1 Introduction

Iron-sulfur proteins are ubiquitous in living organisms. The name "ferredoxin" (Fd) was used for the first time in the early sixties [1], when the electron-transfer function carried out by these proteins was recognized [2]. Recently, an electron transfer physiological role in bacterial photosynthesis has finally been demonstrated also for a smaller class of high potential iron-sulfur proteins (HiPIP) [3].

These proteins have attracted very large attention by researchers involved in understanding why Nature has used iron, and in particular Fe–S polymetallic centers, for electron transfer proteins, and in understanding the factors determining the reduction potentials and electron transfer rates. Some of these questions are addressed by Da Silva and Williams [4], whereas electron transfer processes have been recently reviewed by Gray [5].

Several other enzymatic functions, in which the iron-sulfur cluster takes up the role of a catalyst or a regulator in molecular biotransformations, have been recently discovered [6–11].

Iron-sulfur proteins are characterized by the presence of polymetallic systems containing sulfide ions, in which the iron ions have variable oxidation states. The tetrahedral coordination of high spin iron is generally achieved with four sulfur donors. We will limit our discussion to systems containing FeS_4 coordination units, because we feel that a thorough understanding of these systems is now at hand. This understanding is a necessary basis for tackling the systems with nitrogen and oxygen donors substituting cysteine sulfur, which have the additional complication of strongly dissymmetric iron sites.

The simplest Fe–S cluster protein in constituted by two iron ions bridged by two sulfide ions (Fig. 1B) [12–16]. The oxidized protein contains two iron(III) ions, whereas the reduced protein contains one iron(III) and one iron(II) ion [17].

The other most common polymetallic system is constituted by four iron ions and four sulfide ions placed at the vertices of a cubane-type structure, i.e. with eight atoms at the vertices of a cube (Fig. 1D) [18–36]. In Fe–S cubanes the vertices show alternation of compression and elongation. The iron ions are at the compressed vertices, and the sulfide ions, bridging three metal ions, at the elongated vertices. The formal oxidation numbers of the iron ions can be $[2Fe^{3+}, 2Fe^{2+}]$ or $[1Fe^{3+}, 3Fe^{2+}]$ in low potential ferredoxins. Alternatively, the oxidation numbers of the iron ions in high potential ferredoxins can be $[3Fe^{3+}, 1Fe^{2+}]$, or $[2Fe^{3+}, 2Fe^{2+}]$. These oxidation states were proposed by Carter et al. [20] who, for the first time, explained the differences in redox potentials between low- and high-potential ferredoxin with what has been since known as the "three state hypothesis". The possibility of achieving these three oxidation states in $[Fe_4S_4]$ centers was then confirmed on model compounds by electrochemical experiments [37].

There is a class of proteins containing a $[Fe_3S_4]$ center, in which one iron is missing from the $[Fe_4S_4]$ core (Fig. 1C). Three sulfide ions bridge two iron ions each, while the fourth sulfide bridges three iron ions [26, 29, 38–40]. They can

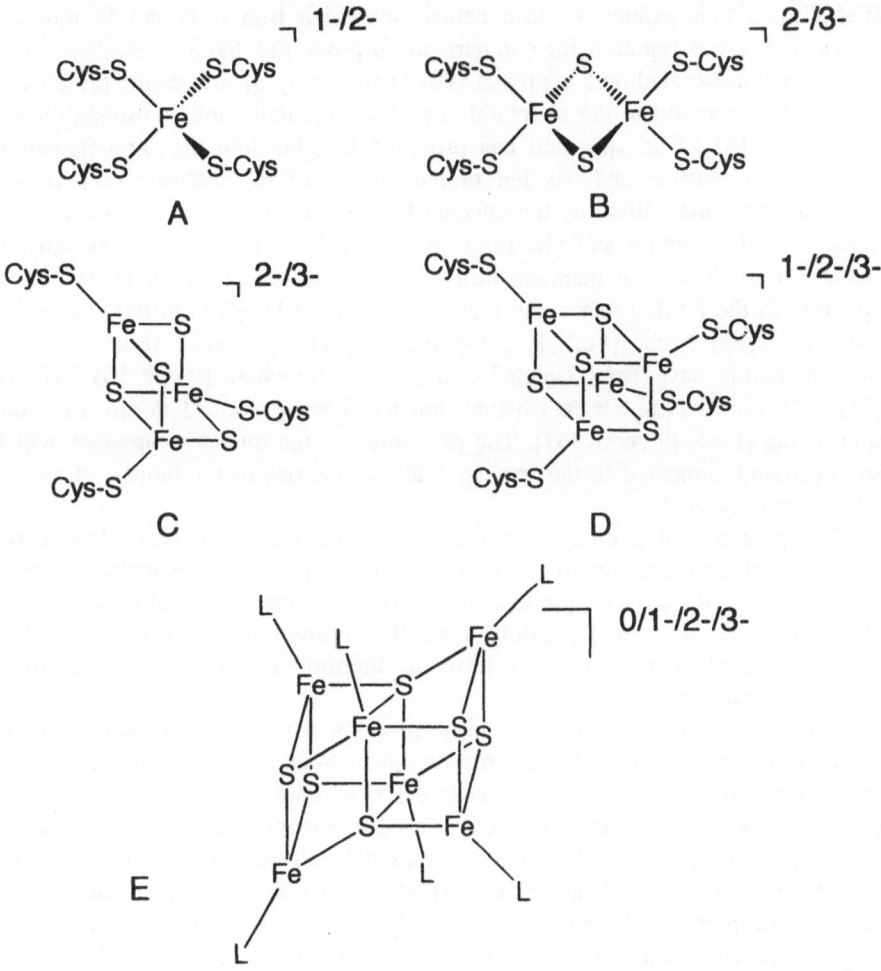

Fig. 1. Schematic structures of iron-sulfur centers found in proteins. L in 6-Fe clusters indicates that the donor atom is not necessarily a Cys sulfur

contain $3Fe^{3+}$ ions or $2Fe^{3+}$ and $1Fe^{2+}$ ion in the oxidized and reduced forms, respectively [41].

Recently, $[Fe_6S_6]$ polymetallic centers (Fig. 1E) have been discovered in proteins [42–45]. The prismane-like structure, proposed on the basis of the fact that the EPR spectra of the isolated proteins are very similar to those observed for the structurally characterized synthetic clusters containing such core [46–52], is that of two superimposed six-membered crowns with three alternating iron and sulfur ions each. The formal oxidation states observed in proteins range from $[6Fe^{3+}]$ to $[3Fe^{3+}, 3Fe^{2+}]$. It has been suggested that two ligands are non-cysteine type, and therefore they will not be treated further here.

In Fig. 1 the schematic structures of the polymetallic cores found in iron-sulfur proteins are summarized. In this figure, the rubredoxin-type, $[Fe(SCys)_4]^{1-}/$

$[Fe(SCys)_4]^{2-}$ monometallic iron center, in which iron is bound to four cysteines, Fig. 1A is reported for comparison purposes [53–59].

The synthetic analogue approach [60] to the study of iron-sulfur proteins has been very successful in the structural and electronic definition of iron-sulfur sites in proteins [61]. The approach has provided insights into the (non-enzymatic) assembly of clusters and has led to elucidation of the *intrinsic* properties of these clusters, unmodified by the effects of protein environment. By these means, *symmetrized* structural and electronic properties have been elucidated. Any departure from these is a manifestation of protein structure which is particularly apparent in the NMR spectra and redox potentials of $[Fe_4S_4]$ clusters. In models, the $[Fe]^{3+}/[Fe]^{2+}$, the $[Fe_2S_2]^{2+}$, and the $[Fe_4S_4]^{1+}/[Fe_4S_4]^{2+}/[Fe_4S_4]^{3+}$ series of compounds have been isolated and structurally characterized [61,62]. The $[Fe_2S_2]^{1+}$ core has not been isolated, but has been generated in situ by reduction of the $[Fe_2S_2]^{2+}$ core [63]. The properties of the model compounds will be surveyed and compared to the corresponding properties of the biological systems when appropriate.

The presence of a number of iron and sulfur atoms in these clusters provides exceptional stability to the system. This is probably the main reason for the spreading of this cofactor in Nature. The soft nature of sulfur also allows charge delocalization, and possibly a wealth of low-lying molecular orbitals to be obtained, which may play a role in the thermodynamics and kinetics of the electron transfer processes.

Some recent reviews on iron-sulfur proteins have been published, and the reader is referred to them for a more complete treatment of the general aspects of the topic [64–69]. Here, we discuss the electronic structure of such clusters (polymetallic systems) as it appears from Mössbauer and magnetic resonance techniques. The problem of localized versus delocalized valence will be tackled, and the contributions of the theoretical tools toward the understanding of the magnetic properties will be analyzed. Finally, the role of the protein environment will be discussed with respect to the valence distribution, the establishment of the redox potential, and the electron transfer processes.

2 Oxidized Two Iron-Two Sulfur Ferredoxins $[Fe_2S_2]^{2+}$

In $[Fe_2S_2]$ proteins, the sulfur bridge provides a means for magnetic interaction between the two high spin, $S = \frac{5}{2}$, Fe^{3+} ions. The nature and the extent of the magnetic coupling has been largely discussed in the theoretical inorganic chemistry community. We can say here that a pragmatic approach is that of describing the interaction through a perturbative Heisenberg Hamiltonian [70]:

$$H = JS_1 \cdot S_2 \tag{1}$$

where S_i and S_j are the localized spin operators for the individual iron ions, and J is the magnetic exchange coupling constant. The latter is mainly due to a superexchange mechanism in which the singly occupied d-orbitals of the iron ions interact through the paired electrons of the sulfur orbitals. Its value depends on the Fe–S distance and on the Fe–S–Fe angle [71,72]. It is considered isotropic, other anisotropic contributions [73] being neglected.

Equation (1) provides eigenstates characterized by S' ranging from $|S_1 + S_2|$ to $|S_1 - S_2|$, with energy values given by

$$E_i = \tfrac{1}{2}JS'_i(S'_i + 1) \tag{2}$$

The ground state is characterized by $S' = |S_1 - S_2| = 0$ because J is positive, and the system is said to be antiferromagnetically coupled.

In the case of $[Fe_2S_2]^{2+}$ oxidized proteins from spinach and S. lividus, the proposed J values range between 290 and 370 cm^{-1} [74,75], the former value being in better agreement with the value obtained for a synthetic analogue [76]. The ground state is then $S' = 0$; however, some paramagnetism arises from the population of the excited states. The magnetic susceptibility is given by the Van Vleck equation that, in the case of $g\mu_B B \ll kT$, can be expressed as [77]:

$$\chi = \frac{g^2\mu_B^2}{3kT}\sum_i S'_i(S'_i + 1)\frac{(2S'_i + 1)\exp(-E_i/kT)}{\sum_i (2S'_i + 1)\exp(-E_i/kT)} \tag{3}$$

where μ_B is the Bohr magneton, k is the Boltzmann constant, and the E_i values, according to Eq. (2), are a function of J. Therefore magnetic susceptibility measurements provide values of J.

The proton NMR spectra show broad, badly resolved signals corresponding to the β-CH$_2$ protons of the cysteines, at about 35 ppm (Fig. 2A) [78–80]. The lack of resolution is due to the long electronic relaxation times of the $S = \tfrac{5}{2}$ manifold [81]. The shift from the diamagnetic value is called the hyperfine shift: in this case it is mainly due to the presence of unpaired spin density on the proton and is said to be contact in origin [12,79,82]. The contribution of a metal j to the contact shift of a proton is given by [79,82]:

$$\left(\frac{\Delta v}{v_0}\right)^{con}_j = \frac{2\pi g\mu_B}{3g_N kT}\cdot\frac{A_j}{h}\cdot\sum_i C_{ji}S'_i(S'_i + 1)\frac{(2S'_i + 1)\exp(-E_i/kT)}{\sum_i (2S'_i + 1)\exp(-E_i/kT)} \tag{4}$$

Here A_j is the hyperfine coupling for the $S = \tfrac{5}{2}$ spin function and the proton nucleus in the case of an uncoupled system. It is related to the amount of unpaired spin density on the proton. The C_{ji} coefficients are given by:

$$C_{1i} = \frac{\langle S_{1z}\rangle_i}{\langle S'_z\rangle_i} \quad C_{2i} = \frac{\langle S_{2z}\rangle_i}{\langle S'_z\rangle_i} \tag{5}$$

where $\langle S_z\rangle$ is the expectation value of S_z. They reflect the contribution of S_1 and S_2 to the total S' of each i level. In the case of homonuclear dimetallic systems, $C_{1i} = C_{2i} = \tfrac{1}{2}$ [83]. Note the similarity between Eqs. (3) and (4), which tells us that the contact shift of protons sensing each individual iron ion is related to

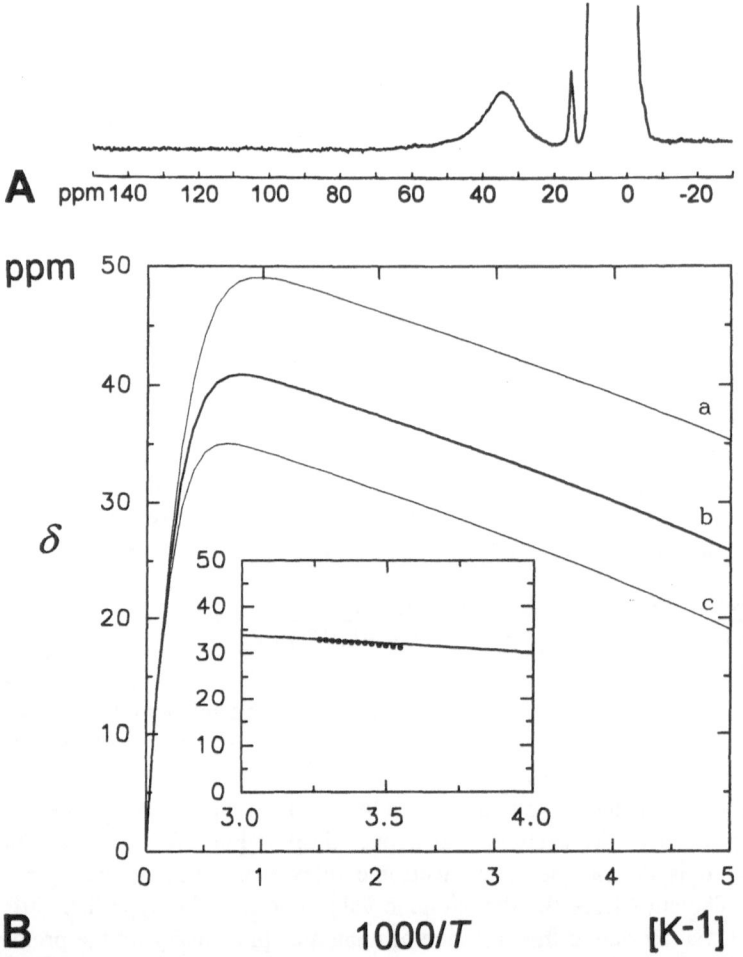

Fig. 2 A. 303 K, 200 MHz ^{1}H NMR spectrum of oxidized $[Fe_2S_2]^{2+}$ ferredoxin from spinach [79, 84]; **B.** Theoretical temperature dependence of the hyperfine shift of β-CH_2 protons of oxidized $[Fe_2S_2]^{2+}$ spinach ferredoxin, calculated using Eq. (14), $J = 250$ (*a*), 290 (*b*), 350 (*c*) cm^{-1}, and $A/h = 1.8$ MHz. In the *inset*, the experimentally observed behavior [84] (*filled circles*) is compared with the theoretical curve (*b*)

the magnetic susceptibility through the coefficients C_{ji}. Therefore, the hyperfine shift and its temperature dependence provide an independent tool to estimate J and to have information on the population of the energy level ladder.

The observed temperature dependence of the NMR hyperfine shifts display an antiCurie behavior, i.e. the shifts increase with increasing temperature because they reflect the population of the excited, magnetically active, levels. Figure 2B shows that the experimental NMR data of the protein from spinach [84] are well reproduced by a theoretical curve obtained from Eq. (4) using $A/h = 1.8$ MHz and E_i obtained from Eq. (2) using the estimated $J = 290$ cm^{-1} [74–76]. The above value for A/h is consistent with experimental observations: $A/h \cong 1$

MHz can be estimated for β-CH_2 protons of cysteines coordinated to iron(II) in rubredoxin and models [85, 86], and A/h values ranging from 1 to 3 MHz have been estimated from proton ENDOR data on a $[Fe_4S_4]^{3+}$ model (see Sect. 6) [87].

The Mössbauer spectra at 4.2 K are characterized by two slightly inequivalent quadrupole pairs centered at 0.20–0.30 mm/s, (as typical for Fe^{3+}) (Table 1) and by the absence of an observable hyperfine field, because of the diamagnetic ground state. Quadrupolar splitting values of 0.6–0.7 mm/s (Refs. in Table 1)

Table 1. Mössbauer parameters for iron-sulfur proteins

Cluster Source	Iron-Sulfur Core	Valence	Fe sites	δ (mm/s) a, b	T (K)	Spin	Ref.
C. pasteurianum Rubredoxin	Fe^{3+}	Fe^{3+}	1	0.32	4.2	5/2	88, 89
C. pasteurianum Rubredoxin	Fe^{2+}	Fe^{2+}	1	0.70	4.2	2	88, 89
Chloropseudomonas ethylica Rubredoxin	Fe^{2+}	Fe^{2+}	1	0.65	77	2	90
D. gigas Rubredoxin	Fe^{2+}	Fe^{2+}	1	0.71	4.2	2	91
D. gigas Desulforedoxin	Fe^{3+}	Fe^{3+}	1	0.25	4.2	5/2	91
D. gigas Desulforedoxin	Fe^{2+}	Fe^{2+}	1	0.70	4.2	2	91
$[Fe(SPh)_4]^{2-}$	Fe^{2+}	Fe^{2+}	1	0.66	4.2	2	92, 93
$[Fe(S_2\text{-}o\text{-xyl})_2]^{1-}$	Fe^{3+}	Fe^{3+}	1	0.25	4.2	5/2	63, 94, 95
$[Fe(S_2\text{-}o\text{-xyl})_2]^{2-}$	Fe^{2+}	Fe^{2+}	1	0.73	4.2	2	63, 94, 95
Synechococcus lividus Fd	$[Fe_2S_2]^{2+}$	$2Fe^{3+}$	2	0.22	4.2	0	75
Synechococcus lividus Fd	$[Fe_2S_2]^{1+}$	$1Fe^{3+}$	1	0.20	250	1/2	75
		$1Fe^{2+}$	1	0.48			
Scenedemus Fd	$[Fe_2S_2]^{2+}$	$2Fe^{3+}$	2	0.20	195	0	96
Scenedemus Fd	$[Fe_2S_2]^{1+}$	$1Fe^{3+}$	1	0.22	195	1/2	96
		$1Fe^{2+}$	1	0.56			
Spinach Fd	$[Fe_2S_2]^{2+}$	$2Fe^{3+}$	2	0.27	4.2	0	17
Spinach Fd	$[Fe_2S_2]^{1+}$	$1Fe^{3+}$	1	0.29	250	1/2	17
		$1Fe^{2+}$	1	0.56			
Parsley Fd	$[Fe_2S_2]^{2+}$	$2Fe^{3+}$	2	0.28	4.2	0	17
Parsley Fd	$[Fe_2S_2]^{1+}$	$1Fe^{3+}$	1	0.29	250	1/2	17
		$1Fe^{2+}$	1	0.58			
Euglena gracilis Fd	$[Fe_2S_2]^{2+}$	$2Fe^{3+}$	2	0.22	4.2	0	97
Euglena gracilis Fd	$[Fe_2S_2]^{1+}$	$1Fe^{3+}$	1	0.25	77	1/2	97
		$1Fe^{2+}$	1	0.70			
C. pasteurianum Fd	$[Fe_2S_2]^{2+}$	$2Fe^{3+}$	2	0.28	4.2	0	17
Putidaredoxin	$[Fe_2S_2]^{2+}$	$2Fe^{3+}$	2	0.18	77	0	98
Putidaredoxin	$[Fe_2S_2]^{1+}$	$1Fe^{3+}$	1	0.35	4.2	1/2	98
		$1Fe^{2+}$	1	0.65			
Adrenodoxin	$[Fe_2S_2]^{2+}$	$2Fe^{3+}$	2	0.27	4.2	0	17, 99
Adrenodoxin	$[Fe_2S_2]^{1+}$	$1Fe^{3+}$	1	0.28	250	1/2	17
		$1Fe^{2+}$	1	0.54			

Table 1. continued

Cluster Source	Iron-Sulfur Core	Valence	Fe sites	δ (mm/s) a, b	T (K)	Spin	Ref.
P. putida Benzene Dioxygenase	$[Fe_2S_2]^{2+}$	$1Fe^{3+}$ $1Fe^{3+}$	1 1	0.23 0.33	77	0	100
P. putida Benzene Dioxygenase	$[Fe_2S_2]^{1+}$	$1Fe^{3+}$ $1Fe^{2+}$	1 1	0.25 0.68	195	1/2	100
T. thermophilus Rieske Fd	$[Fe_2S_2]^{2+}$	$2Fe^{3+}$	1 1	0.24 0.32	4.2	0	101
T. thermophilus Rieske Fd	$[Fe_2S_2]^{1+}$	$1Fe^{3+}$ $1Fe^{2+}$	1 1	0.31 0.74	4.2	1/2	101
Pseudomonas putida 4-methoxy-O-demethylase	$[Fe_2S_2]^{1+}$	$1Fe^{3+}$ $1Fe^{2}$	1 1	0.25 0.70	150	1/2	102
$[Fe_2S_2(S_2\text{-}o\text{-xyl})_2]^{2-}$	$[Fe_2S_2]^{2+}$	$2Fe^{3+}$	2	0.31	4.2	0	63
$[Fe_2S_2(S_2\text{-}o\text{-xyl})_2]^{3-}$	$[Fe_2S_2]^{1+}$	$1Fe^{3+}$ $1Fe^{2+}$	1 1	0.31 0.72	4.2	1/2	63
D. gigas Fd II	$[Fe_3S_4]^{1+}$	$3Fe^{3+}$	3	0.27	77	1/2	41, 103
D. gigas Fd II	$[Fe_3S_4]^{0}$	Fe^{3+} $Fe^{2.5+}-Fe^{2.5+}$	1 2	0.32 0.46	4.2	2	41, 103
A. vinelandii Fd I	$[Fe_3S_4]^{1+}$	$3Fe^{3+}$	3	0.27	77	1/2	104
A. vinelandii Fd I	$[Fe_3S_4]^{0}$	Fe^{3+} $Fe^{2.5+}-Fe^{2.5+}$	1 2	0.29 0.47	4.2	2	104
Aconitase (beef heart)	$[Fe_3S_4]^{1+}$	$3Fe^{3+}$	3	0.27	4.2	1/2	105, 106, 107
Aconitase (beef heart)	$[Fe_3S_4]^{0}$	Fe^{3+} $Fe^{2.5+}$ $Fe^{2.5+}$	1 1 1	0.30 0.47 0.44	4.2	2	105, 106
T. thermophilus Fd	$[Fe_3S_4]^{1+}$	$3Fe^{3+}$	3	0.28	50	1/2	108
Pyrococcus furiosus Fd	$[Fe_3S_4]^{0}$	$1Fe^{3+}$ $2Fe^{2.5+}$	1 2	0.30 0.47	4.2	2	109
$[Fe_4S_4(LS_3)(t\text{-BuNC})_3]^{1-}c$	$[Fe_4S_4]^{2+}$	Fe^{2+} (low spin) Fe^{3+} $Fe^{2.5+}$ $Fe^{2.5+}$	1 1 1 1	0.20 0.34 0.46 0.47	4.2	2	110, 111
C. vinosum Fd	$[Fe_4S_4]^{2+}$	$2Fe^{3+}, 2Fe^{2+}$	4	0.41	77	0	112
C. vinosum Fd	$[Fe_4S_4]^{1+}$	$1Fe^{3+}, 3Fe^{2+}$	4	0.54	77	1/2	112
B. stearothermophilus Fd	$[Fe_4S_4]^{2+}$	$2Fe^{3+}, 2Fe^{2+}$	4	0.42	4.2	0	113, 114, 115
B. stearothermophilus Fd	$[Fe_4S_4]^{1+}$	$1Fe^{3+}, 3Fe^{2+}$	2 2	0.58 0.50	4.2	1/2	113, 114, 115
C. pasteurianum Fd	$[Fe_4S_4]^{2+}$	$2Fe^{3+}, 2Fe^{2+}$	4	0.44	4.2	0	116, 117, 118
C. pasteurianum Fd	$[Fe_4S_4]^{1+}$	$1Fe^{3+}, 3Fe^{2+}$	4	0.58	4.2	1/2	116, 117, 118

Table 1. continued

Cluster Source	Iron-Sulfur Core	Valence	Fe sites	δ (mm/s) a, b	T (K)	Spin	Ref.
A. vinelandii Fd I	$[Fe_4S_4]^{2+}$	$2Fe^{3+}$, $2Fe^{2+}$	4	0.45	4.2	0	104
T. thermophilus Fd	$[Fe_4S_4]^{2+}$	$2Fe^{3+}$, $2Fe^{2+}$	1 3	0.43 0.45	4.2	0	108
D. gigas Fd II	$[Fe_4S_4]^{2+}$	$2Fe^{3+}$, $2Fe^{2+}$	1 3	0.41 0.45	4.2	0	119
Aconitase (beef heart)	$[Fe_4S_4]^{2+}$	$1Fe^{2.5+}$, $3Fe^{2.5+}$	1 3	0.46 0.47	4.2	0	120
C. vinosum HiPIP	$[Fe_4S_4]^{3+}$	$3Fe^{3+}$, $1Fe^{2+}$	2 2	0.29 0.40	4.2	1/2	121, 122
C. vinosum HiPIP	$[Fe_4S_4]^{2+}$	$2Fe^{3+}$, $2Fe^{2+}$	4	0.43	4.2	0	121, 122
C. vinosum HiPIP	$[Fe_4S_4]^{1+}$	$1Fe^{3+}$, $3Fe^{2+}$	4	0.59	77	1/2	123
E. halophila HiPIP II	$[Fe_4S_4]^{3+}$	$3Fe^{3+}$, $1Fe^{2+}$	2 2	0.27 0.37	77	1/2	124
$[Fe_4S_4(S\text{-}2,4,6\text{-}i\text{-}Pr_3C_6H_2)_4]^{1-}$	$[Fe_4S_4]^{3+}$	$3Fe^{3+}$, $1Fe^{2+}$	2 2	0.34 0.40	4.2	1/2	125
$[Fe_4S_4(S\text{-}2,4,6\text{-}i\text{-}Pr_3C_6H_2)_4]^{2-}$	$[Fe_4S_4]^{2+}$	$2Fe^{3+}$, $2Fe^{2+}$	2 2	0.47 0.48	4.2	0	125
A. vinelandii Nitrogenase Iron Protein	$[Fe_4S_4]^{2+}$	$2Fe^{3+}$, $2Fe^{2+}$	1 3	0.44 0.45	4.2	0	126
A. vinelandii Nitrogenase Iron Protein in Ethylene Glycol	$[Fe_4S_4]^{1+}$	$1Fe^{3+}$, $3Fe^{2+}$	2 2	0.53 0.59	50	1/2	126
A. vinelandii Nitrogenase Iron Protein in Urea	$[Fe_4S_4]^{1+}$	$1Fe^{3+}$, $3Fe^{2+}$	2 2	0.54 0.54	50	$\frac{3}{2}$	126
B. subtilis amidotrasferase	$[Fe_4S_4]^{2+}$	$2Fe^{3+}$, $2Fe^{2+}$	4	0.43	100	0	10
B. subtilis amidotrasferase	$[Fe_4S_4]^{1+}$	$1Fe^{3+}$, $3Fe^{2+}$	4	0.56	100	$\geq \frac{3}{2}$	10
$[Fe_4S_4(SEt)_4]^{2-}$	$[Fe_4S_4]^{2+}$	$2Fe^{3+}$, $2Fe^{2+}$	4	0.46	4.2	0	127
$[Fe_4S_4(S\text{-}t\text{-}Bu)_4]^{2-}$	$[Fe_4S_4]^{2+}$	$2Fe^{3+}$, $2Fe^{2+}$	4	0.47	4.2	0	127
$[Fe_4S_4(SCH_2Ph)_4]^{2-}$	$[Fe_4S_4]^{2+}$	$2Fe^{3+}$, $2Fe^{2+}$	4	0.46	4.2	0	127, 128
$[Fe_4S_4(SCH_2Ph)_4]^{3-}$	$[Fe_4S_4]^{1+}$	$1Fe^{3+}$, $3Fe^{2+}$	2 2	0.60 0.60	4.2	1/2	129
$[Fe_4S_4(SPh)_4]^{2-}$	$[Fe_4S_4]^{2+}$	$2Fe^{3+}$, $2Fe^{2+}$	4	0.47	4.2	0	127
$[Fe_4S_4(SPh)_4]^{3-}$	$[Fe_4S_4]^{1+}$	$1Fe^{3+}$, $3Fe^{2+}$	2 2	0.57 0.64	4.2	1/2	129

[a] All isomer shifts are referenced to, or are corrected to refer to, iron metal at room temperature
[b] Errors in the isomer shift are in the range 0.005–0.03 mm/s
[c] LS_3 = 1,3,5-tris((4,6-dimethyl-3-mercaptophenyl)thio)-2,4,6-tris(p-tolylthio)benzene(3-)

indicate a relatively symmetric environment, consistent with the pseudotetrahedral S_4 coordination polyhedron [17]. Synthetic, structurally centrosymmetric models of the oxidized $[Fe_2S_2]^{2+}$ proteins have been prepared [63, 76, 130–132]. Their electronic features are very similar to the corresponding proteins, although the Mössbauer spectra reveal a higher symmetry than that encountered in proteins [76]. This is also shown by resonance Raman spectroscopy [133] and, more directly, by X-ray crystallography [13–16].

In a way, despite the magnetic coupling which prevents deep investigations of the electronic structure for each metal ion, a clear picture of the dimetallic system emerges.

3 Cores Containing $2Fe^{3+}$ and $2Fe^{2+}$ Ions

Such cores contain metal ions that are only formally in different oxidation states, but actually contain four iron ions which appear equivalent by Mössbauer spectroscopy. The isomer shifts (Table 1) are intermediate between those found for systems containing Fe^{2+} and Fe^{3+} [134, 135], and are consistent with four iron ions in the oxidation state of 2.5+. The polymetallic center contains an even number of electrons and the ground state is diamagnetic ($S' = 0$) [136]. Diamagnetism can be obtained by extension of the Hamiltonian in Eq. (1) to:

$$H = \sum_{i \neq j} J_{ij} \mathbf{S}_i \cdot \mathbf{S}_j \tag{6}$$

An equivalent way which provides an analytical solution for the eigenvalues is:

$$H = J \sum_{i \neq j} \mathbf{S}_i \cdot \mathbf{S}_j + \Delta J_{12} \mathbf{S}_1 \cdot \mathbf{S}_2 + \Delta J_{34} \mathbf{S}_3 \cdot \mathbf{S}_4 \tag{7}$$

with relative energies given by:

$$E(S_{12}, S_{34}, S') = \tfrac{1}{2}[JS'(S' + 1) + \Delta J_{12} S_{12}(S_{12} + 1) + \Delta J_{34} S_{34}(S_{34} + 1)] \tag{8}$$

where $J = J_{13} = J_{14} = J_{23} = J_{24}$, $\Delta J_{12} = J_{12} - J$, $\Delta J_{34} = J_{34} - J$, S' is the total spin of the cluster, and S_{12} and S_{34} are subspin quantum numbers relative to the (S_1-S_2) and (S_3-S_4) spin pairs. Of course the J values can be all equal. As in the case of oxidized $[Fe_2S_2]^{2+}$, a ladder of levels with $S' > 0$ above the $S' = 0$ ground state is obtained.

The ^1H NMR spectra show some hyperfine shifted signals due to the residual paramagnetism, as discussed in Sect. 2 for the case of $[Fe_2S_2]^{2+}$ proteins. The extent of the hyperfine coupling for the SCH_2R protons of the iron-bound thiolate falls in the same range for models [86, 137], reduced HiPIPs [138–151], and oxidized ferredoxins [152–159], thus providing evidence for the similarity of the coordination bond. Typical protein spectra are shown in Fig. 3. The temperature dependence of the shifts of all hyperfine shifted signals is of antiCurie type. Using Hamiltonian (7) with $\Delta J_{ij} = 0$ and whatever J values, Curie and antiCurie behaviors are predicted, contrary to observation. On the other hand, the experimental antiCurie temperature dependence of the signals can be reproduced with $\Delta J_{12} \cong \Delta J_{34} < 0$, J being about 150–200 cm^{-1} [154, 157].

The results from Mössbauer and NMR spectroscopy indicate that the description based on localized unpaired electrons on the single ions is poor, and that some sort of electron delocalization occurs. This phenomenon has been theoretically studied on dinuclear [160–164], trinuclear [103], and tetranuclear centers [165–169]. The requirement of delocalization between a high spin iron(III)

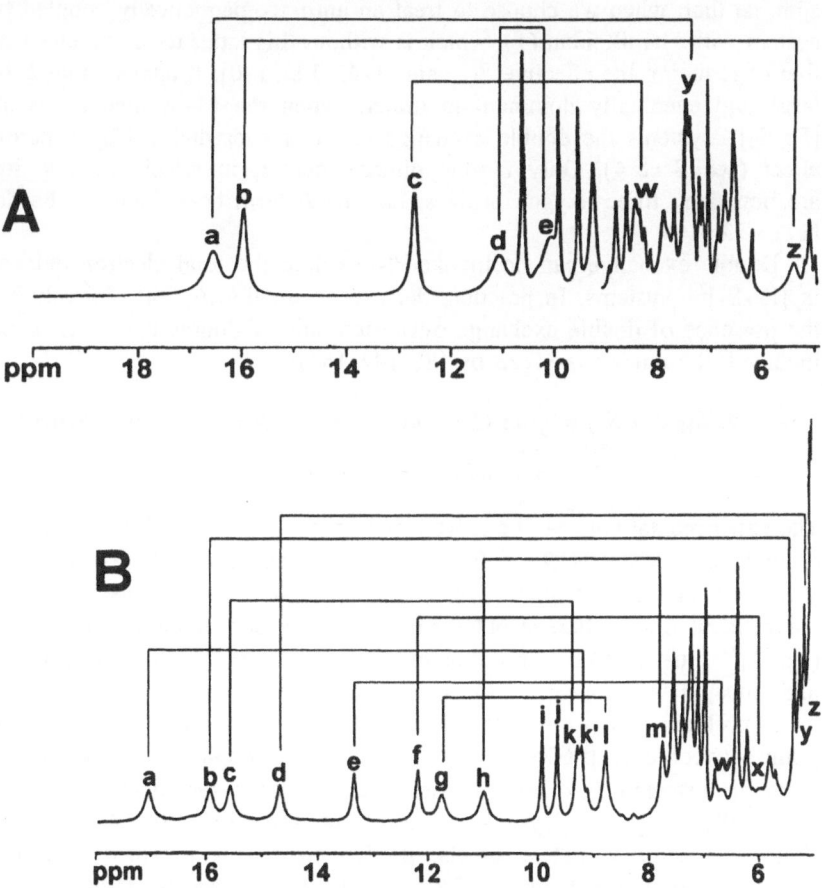

Fig. 3 A. 300 K, 600 MHz ^{1}H NMR spectrum of reduced [Fe$_4$S$_4$]$^{2+}$ HiPIP from *C. vinosum*. The cysteine β-CH$_2$ geminal connectivities are indicated [142]; B. 300 K, 600 MHz ^{1}H NMR spectrum of oxidized 2[Fe$_4$S$_4$]$^{2+}$ ferredoxin from *C. pasteurianum*. The cysteine β-CH$_2$ geminal connectivities are indicated [157]

and a high spin iron(II) can be formally taken in consideration modifying the spin Hamiltonian (7) as [103, 170]

$$H = [J^1 \mathbf{S}_1 \cdot {}^1\mathbf{S}_2] \cdot \mathbf{O}_1 + [J^2 \mathbf{S}_1 \cdot {}^2\mathbf{S}_2] \cdot \mathbf{O}_2 + B_{12} \mathbf{V}_{12} \mathbf{T}_{12} \qquad (9)$$

where $\mathbf{O}_i$ are the occupation operators which keep track of the position of the extra electron, $\mathbf{T}_{12}$ is the transfer operator between sites 1 and 2, $\mathbf{V}_{12}$ is an operator producing as eigenvalues $(S' + \frac{1}{2})$, ${}^j S_i$ are the spin operators S_i when the extra electron is on site j, and B_{12} is the scalar value proportional to the extent of delocalization. The energies are given by [103, 170]

$$E = \tfrac{1}{2} J S'(S' + 1) \pm B_{12}(S' + \tfrac{1}{2}) \qquad (10)$$

Historically, B is called double exchange [160, 161]. B is a ferromagnetic contribution because the electron moves from one ion to the other retaining its spin, so that, when we choose to treat an antiferromagnetically coupled two-spin system with Hamiltonian (1), which is without this latter term, the effect of B_{12} is that of reducing the effective J_{12} value [142, 143, 170]. It appears that B becomes large and eventually dominant in dimers when the M–M distance is short. In $[Fe_2S_2]^+$ systems the double exchange effect is overwhelmed by superexchange effect (see Sect. 4). Only in one dimeric center, in which the two iron ions are bound by three oxygen bridges, has the B term been found to be dominant [171, 172a].

Double exchange can be invoked to explain the total electron delocalization in $[Fe_4S_4]^{2+}$ systems. In practice, the use of smaller J_{12} and J_{34} values mimics the presence of double exchange over each pair. If double exchange is explicitly included, the energy is given by [82, 142, 154]:

$$E(S_{12}, S_{34}, S') = \tfrac{1}{2}[JS'(S' + 1) + \Delta J_{12}S_{12}(S_{12} + 1) + \Delta J_{34}S_{34}(S_{34} + 1)]$$
$$\pm B_{12}(S_{12} + \tfrac{1}{2}) \pm B_{34}(S_{34} + \tfrac{1}{2}) \qquad (11)$$

The experimental data can be reproduced equally well with $\Delta J_{ij} < 0$ and $B_{ij} = 0$, or with $\Delta J_{ij} = 0$ and $B_{ij} \neq 0$. Thus there is covariance between ΔJ_{ij} and B_{ij} terms and therefore estimates of their respective values is not possible, unless independent information is obtained. The above considerations indicate that the $[Fe_4S_4]^{2+}$ cluster can be viewed as made by two pairs, S_{12} and S_{34}, with relatively high subspin values (e.g. $\frac{9}{2}$, $\frac{9}{2}$), antiferromagnetically coupled one another.

A problem which arises is the nature of the inequivalence of hyperfine coupling between each β-CH$_2$ proton and the unpaired spin density on the center. A thorough research has been carried out to stereo-specifically and sequence-specifically assign the cysteine protons [144–148, 151, 158, 173–176]. Through NOEs the pairwise nature of the β-CH$_2$ protons is obtained [142, 154, 177]. This has been verified through NOESY and COSY spectra [144–148, 155–159, 173–176], in such a way that the α-CH protons could also be assigned. In order to proceed with the sequence-specific assignment, the TOCSY or COSY fingerprints of amino acidic residues dipolarly connected with the cysteine protons had to be recognized. For the cases in which the X-ray structure is known, the assignment of a large number of protons around the metal ions is possible. In the cases in which the X-ray structure is not available, structural models could be generated through computer graphics methods by starting from X-ray data of homologous proteins [174]. Sophisticated MD approaches have been deviced, which were able to refine the X-ray data in order to make proton-proton distances more consistent with NOE experiments [178]. In every case, the independently generated structure was compared with, and corroborated by, NOE data. The general approach for the sequence-specific assignment exploits either the reduced or the oxidized protein spectra, whose signals are usually connected through EXSY experiments.

Once a β-CH₂ Cys proton is recognized to selectively give NOE with a portion of an assigned residue, by inspection of the structural model, the sequence specific, stereoselective assignment is performed. A longer procedure is that of assigning all the protein residues and solve the solution structure using the NMR data.

The proton hyperfine coupling of a metal-donor β-CH_2 moiety depends on the dihedral angle θ. This angle is defined so that $\theta = 0$ corresponds to the eclipsed orientation of the metal and the β-CH proton of the bound $-SCH_2-$ group, as shown in Fig. 4A. When the proton $1s$ orbital receives spin density through direct overlap with σ metal-donor, the angular dependence of the isotropic shift follows the Karplus [179, 180] relationship:

$$\delta_\sigma = a' \cos^2\theta + b'\cos\theta + c' \tag{12}$$

where a', b' and c' depend on the actual electronic structure. b' and c' are usually small, and are often neglected [181–183]. When on the other hand the spin density on the proton comes from direct overlap of the $1s$ orbital with a donor p_z orbital, the hyperfine shift is instead given by [184–186]:

$$\delta_\pi = a'' \sin^2\theta + c'' . \tag{13}$$

In the present case, both mechanisms can be operative, and the sum of Eqs. (12) and (13) has the form [158]:

$$\delta = a \sin^2\theta + b \cos\theta + c \tag{14}$$

where a can be positive or negative depending on whether the latter or the former mechanism predominates, respectively. In Fig. 4B the experimental hyperfine shift values of β-CH_2 protons of oxidized *C. acidi urici* and *C. pasteurianum* ferrodoxins, and reduced *C. vinosum*, *E. halophila* iso-I, *E. halophila* iso-II, and *E. vacuolata* iso-II HiPIPs are reported, together with a fit of function (14) to the experimental data (solid line). From this fit, a positive value of a is calculated (10.85 ppm), indicating that the π overlap is the dominant mechanism. Furthermore, the hyperfine shifts of Cα follow the same pattern, as also shown in Fig. 4B. It is thus possible that this mechanism dominates in all Fe–S systems [146], even though other effects may introduce large inequivalences among the various protons of the bound cysteine residues, complicating the interpretation.

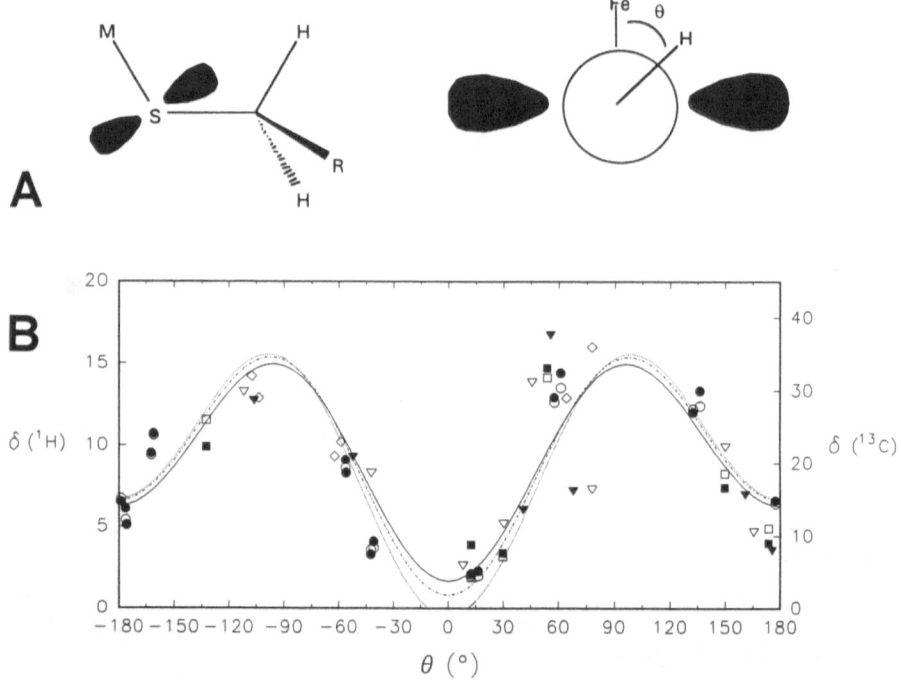

Fig. 4 A. Schematic view of the M-S-C-H dihedral angle θ. A p_z orbital of S, orthogonal to the M-S-C-plane is also shown; **B.** Plot of the hyperfine shifts of cysteine β-CH$_2$ protons (*left hand scale*), and Cα carbons (*right hand scale*), for some proteins containing [Fe$_4$S$_4$]$^{2+}$ clusters, as a function of the dihedral angle θ, Fe-S-Cβ-H and Fe-S-Cβ-Cα, respectively. The hyperfine shifts from different proteins are reported as follows: (●) oxidized *C. acidi urici* ferredoxin; (○) oxidized *C. pasteurianum* ferredoxin; (▽) reduced *C. vinosum* HiPIP; (▼) reduced *E. vacuolata* HiPIP I; (□) reduced *E. halophila* HiPIP I; (■) reduced *E. halophila* HiPIP II; (◇) oxidized *C. acidi urici* Fd Cα. The curves report the fittings of the values on the basis of Eq. (14) (see text); the best fit values obtained for the three parameters are: $a = 10.85, b = -2.29, c = 3.98$ ppm (*solid line*); $a = 11.5, b = -2.9, c = 3.7$ (*dashed line*); $a = 12.2, b = -3.6, c = 3.1$ (*dotted line*)

This analysis shows that, if the effect of the dihedral angles is taken into account, the hyperfine values of all the nuclei of the cysteines bound to the four iron ions of a [Fe$_4$S$_4$]$^{2+}$ core are essentially equal. This is an important result as far as the equivalence of the electronic structure of the iron ions is concerned, which is consistent with the Mössbauer data. Note that the latter are taken at 4.2 K, whereas the NMR data are taken at room temperature. The values $a = 11.5$, $b = -2.9$, and $c = 3.7$ were previously proposed [158] by fitting the *C. acidi urici* data using the angles of the highly homologous *P. aerogenes* Fd (dashed line in Fig. 4B) [21, 24, 133b]. Later, on the basis of the X-ray structure of *C. acidi urici* Fd [33], the parameters were modified as follows: $a = 12.2$, $b = -3.6$, $c = 3.1$ (dotted line in Fig. 4B) [187]. The a, b and c values of 10.85, −2.29, and 3.98, respectively, are thus proposed now as parameters to obtain structural information from the β-CH$_2$ shifts. They are calculated by using the dihedral angles taken from the structure of each protein or of the closest analogue.

In model compounds containing the $[Fe_4S_4]^{2+}$ core, the dihedral angle dependence is averaged out by rotation, fast on the NMR time scale, through the –S–C bond of the thiolate bound to the cluster. The experimental hyperfine shift of ca. 10 ppm [86, 138] is in excellent agreement with the average value of 9.40 ppm calculated from Eq. (14) using the parameters obtained to fit the experimental data for all proteins.

In summary, a clear picture with four equivalent $Fe^{2.5+}$ ions and complete electron delocalization both in proteins and in model compounds is obtained. A modified Heisenberg approach is able to account for the main spectroscopic properties. The hyperfine shift of β-CH_2 Cys protons provide structural information and a detailed mechanism of unpaired spin density transfer.

4 Reduced Two Iron-Two Sulfur Ferredoxins $[Fe_2S_2]^+$

This is an example in which two iron ions are actually inequivalent and conveniently described as one Fe^{3+} and one Fe^{2+} ion. In fact, the Mössbauer spectra indicate that there is one iron(III) and one iron(II), with isomer shifts similar to those of the uncoupled system (Table I). The magnetic coupling described by Eq. (1) accounts for an EPR ground state with $S' = \frac{1}{2}$ and for the intensity of the $S' = \frac{1}{2}$ EPR signal decreasing, and broadening, with increasing temperature [188–192]. $[Fe_2S_2]^+$ cores should allow for double exchange mechanisms. The experimental observations that the valences are localized, that the spin ground state has the smallest multiplicity, and that the two sites are inequivalent, show that double exchange is overwhelmed by antiferromagnetic coupling. Under these conditions, it has been shown [170, 193] that the effect of B is limited to a reduction of the effective J value. Measurements of the J constant for reduced $[Fe_2S_2]$ proteins using either the temperature dependence of the magnetic susceptibility, or of the intensity of the EPR spectra, have yielded the values of 180–220 cm^{-1} for spinach Fd [74, 191], 230 cm^{-1} for *Synecococcus lividus* Fd [75], 170 cm^{-1} for *Spirulina maxima* Fd [190], 180 cm^{-1} for succinate dehydrogenase center S-1 and NADH:UQ reductase center N-1a [189], and 150 cm^{-1} for *Halobacterium halobium* Fd [191]. Higher J values of 540 cm^{-1} for beef adrenal cortex mitochondria Fd [189], and 330 cm^{-1} for pig adrenal glands Fd [191], have been reported. For the spinach and *Synecococcus lividus* proteins, J values for both reduced [74, 75, 191] and oxidized [74, 75] forms are available. The smaller magnetic exchange coupling constant observed in the reduced form can be due partly to the larger ionic radius of iron(II) which makes the Heisenberg exchange mechanism less efficient, and partly to a double exchange contribution. In order to account for electron localization, it has previously been proposed that B must be smaller than $4.5J$ [193], without considering the effect of iron reduction. If this latter effect is taken into consideration, a much lower upper limit can be estimated for B [194]: even if $J_{Fed} = J_{ox}(= 290 \text{ cm}^{-1})$ B can hardly be larger than 150 cm^{-1} [194]. This would be also in agreement with Mössbauer data on a partially delocalized $[Fe_2S_2]^+$ model compound [172b].

ENDOR experiments have allowed the measurement of the principal components of the ^{57}Fe magnetic hyperfine tensors [198]: one iron ion has an almost isotropic A-tensor of magnitude ca. 46 MHz, and the other iron ion has a highly anisotropic A-tensor, with principal values of about 17, 24, and 35 MHz for adrenodoxin and putidaredoxin [198]. While ENDOR does not allow for the determination of the sign of the ^{57}Fe hyperfine constants, magnetic Mössbauer spectra at 4.2 K indicated the presence of two non-equivalent iron sites, one with a negative, slightly anisotropic A-tensor at around -46 MHz, the other with a positive, highly anisotropic A-tensor (ca. $+11$, $+14$, and $+34$ MHz) [17]. ^{57}Fe hyperfine constants are negative for Fe(III) and positive for Fe(II) [12]. It is striking that the signs of the hyperfine coupling are different! This is because at 4.2 K only the ground state is occupied, and in such level the smaller $S = 2$ spin has a different orientation than the larger $S = \frac{5}{2}$ spin because of anitferromagnetic coupling. Therefore the coupling energy between $^{57}Fe(III)$ and $S = \frac{5}{2}$ will be of different sign than that between $^{57}Fe(II)$ and $S = 2$. This comes also from the values of the C_{ji} coefficients for the ground state (see Eq. 5), which are $+\frac{7}{3}$ for $S = \frac{5}{2}$ and $-\frac{4}{3}$ for $S = 2$ [82].

The observed g values for a number of reduced $[Fe_2S_2]$ ferredoxins are reported in Table 2. According to the theory of magnetically coupled systems,

Table 2. EPR parameters for iron-sulfur proteins

Cluster Source	Core	g Values	Spin	Ref.
C. ethylica Rubredoxin	Fe^{3+}	9.4, 4.3	5/2	90
P. oleovorans Rubredoxin	Fe^{3+}	9.42, 4.31	5/2	195
D. gigas Desulforedoxin	Fe^{3+}	7.7, 4.1, 1.8, 5.7	5/2	91
Synechococcus lividus Fd	$[Fe_2S_2]^{1+}$	2.05, 1.96, 1.88	1/2	75
Scenedemus Fd	$[Fe_2S_2]^{1+}$	2.037, 1.944, 1.888	1/2	96
Microcystis flos-aquae Fd	$[Fe_2S_2]^{1+}$	2.05, 1.96, 1.89	1/2	196
Euglena gracilis Fd	$[Fe_2S_2]^{1+}$	2.056, 1.964, 1.896	1/2	197
H. halobium Fd	$[Fe_2S_2]^{1+}$	2.067, 1.982, 1.89	1/2	191
Spinach Fd	$[Fe_2S_2]^{1+}$	2.041, 1.949, 1.884	1/2	96
Parsley Fd	$[Fe_2S_2]^{1+}$	2.03, 1.96, 1.90	1/2	198
Putidaredoxin	$[Fe_2S_2]^{1+}$	2.02, 1.935, 1.93	1/2	198
Adrenodoxin	$[Fe_2S_2]^{1+}$	2.02, 1.935, 1.93	1/2	198
T. thermophilus Rieske Fd	$[Fe_2S_2]^{1+}$	2.02, 1.90, 1.80	1/2	101
Yeast Rieske Fd	$[Fe_2S_2]^{1+}$	2.02, 1.89, 1.81	1/2	199
R. sphaeroides Rieske Fd	$[Fe_2S_2]^{1+}$	2.03, 1.90, 1.81	1/2	200
P. putida Benzene Dioxygenase	$[Fe_2S_2]^{1+}$	2.01, 1.91, 1.75	1/2	201

Table 2. continued

Cluster Source	Core	g Values	Spin	Ref.
Pseudomonas putida 4-methoxy-O-demethylase	$[Fe_2S_2]^{1+}$	2.008, 1.913, 1.72	1/2	202, 203
Pyrazon dioxygenase	$[Fe_2S_2]^{1+}$	2.02, 1.91, 1.79	1/2	204
$[Fe_2S_2(S_2\text{-}o\text{-xyl})_2]^{3-}$	$[Fe_2S_2]^{1+}$	2.01, 1.94, 1.92	1/2	63
D. gigas Fd II	$[Fe_3S_4]^{1+}$	2.02, 2.00, 1.97	1/2	41
Aconitase (beef heart)	$[Fe_3S_4]^{1+}$	2.004, 2.016, 2.024	1/2	107
T. thermophilus Fd	$[Fe_3S_4]^{1+}$	"g = 2.02"	1/2	108
M. barkeri Fd	$[Fe_3S_4]^{1+}$	"g = 2.02"	1/2	205
C. vinosum HiPIP	$[Fe_4S_4]^{3+}$	2.12, 2.04, 2.04 2.088, 2.055, 2.040 2.12, 2.04, 2.02 2.13, 2.07, 2.04	1/2	206, 207
E. halophila HiPIP II	$[Fe_4S_4]^{3+}$	2.078, 2.033, 2.033	1/2	124, 208
R. gelatinosus HiPIP	$[Fe_4S_4]^{3+}$	2.11, 2.03, 2.03	1/2	208
E. vacuolata HiPIP I	$[Fe_4S_4]^{3+}$	2.10, 2.03, 2.03	1/2	150
E. vacuolata HiPIP II	$[Fe_4S_4]^{3+}$	2.10, 2.03, 2.03	1/2	175
C. tepidum HiPIP	$[Fe_4S_4]^{3+}$	2.12, 2.04, 2.04	1/2	209
$[Fe_4S_4(S\text{-}2,4,6\text{-i-}Pr_3C_6H_2)_4]^{1-}$	$[Fe_4S_4]^{3+}$	2.10, 2.07, 2.03	1/2	125
Center I in irradiated $[Fe_4S_4(SCH_2C_6D_5)]^{2-}$	$[Fe_4S_4]^{3+}$	2.142, 2.013, 2.004	1/2	87
Center IV in irradiated $[Fe_4S_4(SCH_2C_6D_5)]^{2-}$	$[Fe_4S_4]^{3+}$	2.070, 2.026, 2.018	1/2	87
C. vinosum HiPIP	$[Fe_4S_4]^{1+}$	2.04, 1.93, 1.93	1/2	112
A. vinelandii Nitrogenase Iron Protein in Ethylene Glycol	$[Fe_4S_4]^{1+}$	2.05, 1.94, 1.88	1/2	126
A. vinelandii Nitrogenase Iron Protein in Urea	$[Fe_4S_4]^{1+}$	5.8, 5.15	$\frac{3}{2}$	126
C. vinosum Fd	$[Fe_4S_4]^{1+}$	2.067, 1.945, 1.894	1/2	196
C. pasteurianum Fd	$[Fe_4S_4]^{1+}$	2.06, 1.94, 1.94	1/2	210
Bacillus polymyxa Fd	$[Fe_4S_4]^{1+}$	2.06, 1.93, 1.88	1/2	211
B. stearothermophilus Fd	$[Fe_4S_4]^{1+}$	2.07, 1.93, 1.89	1/2	114
Micrococcus lactylicus Fd	$[Fe_4S_4]^{1+}$	2.07, 1.94, 1.89	1/2	212
R. rubrum Fd III	$[Fe_4S_4]^{1+}$	2.04, 1.93, 1.93	1/2	213
$[Fe_4S_4(SCH_2Ph)_4]^{3-}$	$[Fe_4S_4]^{1+}$	2.04, 1.93, 1.93	1/2	129
$[Fe_4S_4(SPh)_4]^{3-}$	$[Fe_4S_4]^{1+}$	2.06, 1.93, 1.93	1/2	129

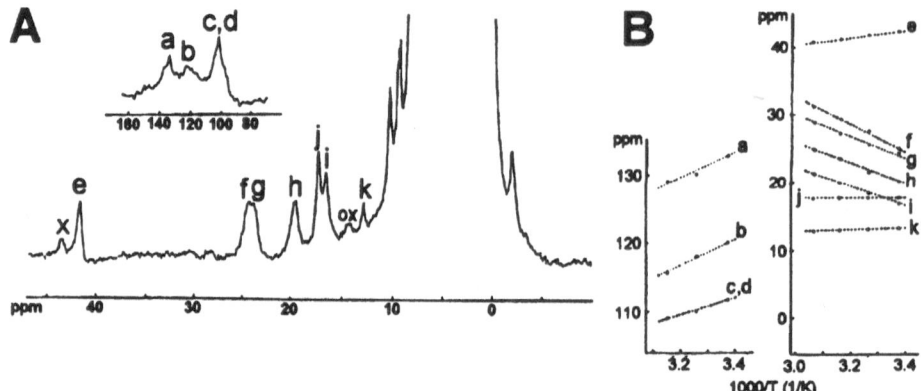

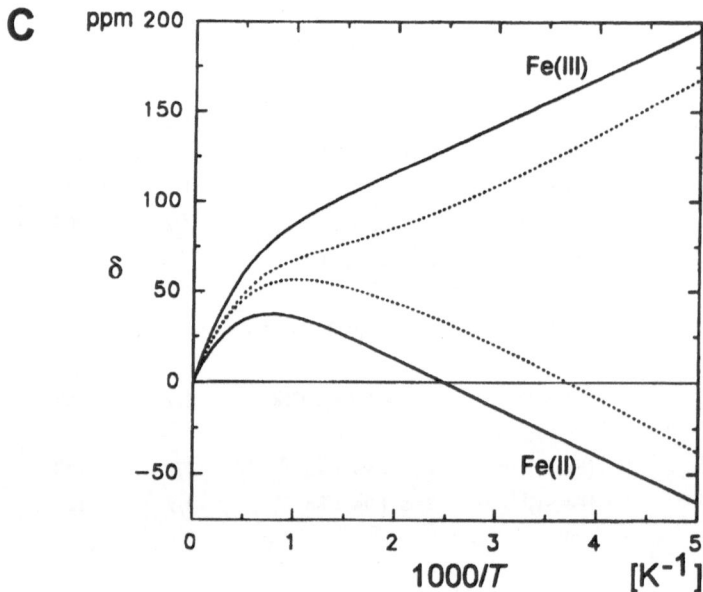

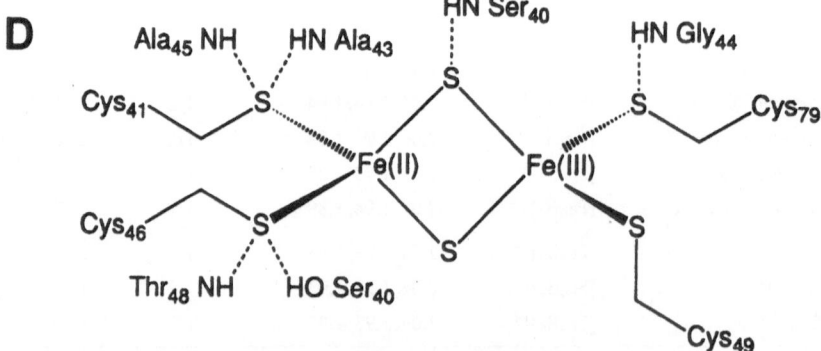

the g values should be a linear combination of those of the uncoupled ions [73, 192, 214–217], which are, however, unknown. It has been calculated, for example, that by taking isotropic $g = 2.019$ for iron(III), by considering a C_{2v} distortion of the Fe^{2+} site, and by assuming that the g_{zz} vector of the Fe^{2+} ion is along the Fe–Fe direction, the resulting g values of the iron(II) ion are $g_x = 2.122$, $g_y = 2.077$, $g_z = 2.002$, and the calculated g values for the coupled system are $g_x = 1.88$, $g_y = 1.94$, $g_z = 2.04$ in good agreement with the experimental results [214]. Note that the average g values of the magnetically coupled $S' = \frac{1}{2}$ state are smaller than 2, as a result of antiferromagnetic coupling. A theory has been proposed [191, 192, 216], which predicts, in good accordance with experiments, that the value of the magnetic coupling constant is larger the larger the symmetry around the ferrous ion.

Mössbauer spectra do not exclude the presence of an equilibrium between two molecular species with different valence distribution, i.e., either iron ion may be reduced. In this respect, the interpretation of the ^1H NMR spectra of proteins containing the $[Fe_2S_2]^+$ core is quite enlightening. It has already been mentioned that high spin iron(III) has long electronic relaxation times which provide broad NMR lines, and that high spin iron(II) is expected to have fast electron relaxation and to provide sharp proton NMR lines [81]. Magnetic coupling is expected, to a first approximation, to decrease the electronic relaxation times of Fe(III), thus making the electronic relaxation times of the two ions similar. This is because the slow relaxing metal ion takes advantage of the relaxation mechanisms of the fast relaxing ion through exchange coupling [81]. As a matter of fact, the linewidths of the protons of the two iron domains are usually different by about one order of magnitude. The ^1H NMR spectrum of reduced spinach Fd is reported in Fig. 5A, and the temperature dependence of the hyperfine shifted signals in Fig. 5B [218]. It is striking that the far-shifted signals have an essentially Curie behavior, while the less downfield-shifted signals have an antiCurie behavior. This can be qualitatively understood by considering that a larger $S = \frac{5}{2}$, associated with Fe(III), forces a smaller spin $S = 2$, associated with Fe(II), to be oriented in an antiparallel fashion in the ground state of this antiferromagnetically coupled system. Therefore, as far as the ground state is concerned, the shifts of the protons sensing the $S = 2$ ion will have opposite sign with respect to that with $S = \frac{5}{2}$. When the temperature is increased, excited levels are populated and the shifts of

---

Fig. 5 A. 297 K, 300 MHz ^1H NMR spectrum of reduced $[Fe_2S_2]^{1+}$ spinach Fd. Signals from residual oxidized protein (ox) and from minor components (x) are indicated [218]; B. Temperature dependence of the ^1H NMR shifts of signals of reduced spinach Fd. The signals are labeled as in A [218]; C. Theoretical temperature dependence of the hyperfine shift of β-CH_2 cysteine protons of reduced $[Fe_2S_2]^{1+}$ spinach Fd (———), calculated using Eq. (4), with J = 200 cm^{-1} and A/h = 1.8 MHz. The analogous curve calculated with Eq. (4), using the same parameters but introducing an equilibrium between two species differing for the location of the extra electron is also shown (-----). In the latter case, the calculated curve is obtained with Δ = 300 cm^{-1} (see text), corresponding to a 20/80 ratio of the two species at room temperature [84]; D. Sequence-specific assignment of cysteine protons in reduced $[Fe_2S_2]^{1+}$ S. platensis Fd [242] and valence-specific assignment of the iron ions

the protons sensing the $S = 2$ iron will move toward the values of the protons sensing the $S = \frac{5}{2}$ iron ion until, in the infinite temperature limit, the protons of the two iron domains will have similar values.

A semiquantitative estimate of the temperature dependence of the shifts is shown in Fig. 5C. Here we have considered that Fe^{2+} has only one $S = 2$ state. Of course this introduces some error because the quintet E ground state of a tetrahedral Fe^{2+} will be split by low symmetry components and the two quintets will experience different J couplings [12]. This approximation will be introduced in all subsequent treatments. From Eq. (4) and Equation (5) the contact shifts can be evaluated by knowing the energy levels from Equation (2). Figure 5C (solid line) shows that, with the J value of 200 cm^{-1}, experimentally observed for the spinach reduced protein [74, 191], it is predicted that the hyperfine shifts of the β-CH$_2$ protons of the cysteines bound to Fe^{3+} are ca. 150 ppm downfield, whereas the corresponding shifts of β-CH$_2$ protons of the cysteines bound to Fe^{2+} are ca. 30 ppm upfield, if the hyperfine coupling constant $A/h = 1.8$ MHz is equal for the two metal ions, and without considering the angular dependence of the type of Eq. (14). That the proton hyperfine coupling constant is similar for the two ions is confirmed also by the similar hyperfine values with ^{57}Fe, for oxidized and reduced rubredoxin [90]. From the comparison between experimental data and expectation, it is concluded that the far hyperfine shifted signals a, b, c, d, and e belong to the four β-CH$_2$ and one α-CH protons of cysteines bound to Fe^{3+} (Fig. 5A). The signals f, g, h, and i are then assigned to the β-CH$_2$ of the cysteines bound to Fe^{2+} (Fig. 5A). The antiCurie behavior of the latter signals is somewhat exaggerated by the calculations, leading to upfield shifted signals. Larger J values decrease the downfield shift of the protons sensing the Fe^{3+} ion and further increase the upfield shift of the protons sensing the Fe^{2+} ion (not shown). This behavior is in qualitative agreement with the observed NMR spectra of reduced bovine adrenodoxin [80], for which a much larger J value was measured [189], and which indeed shows upfield shifted signals.

From these spectra it can be concluded that, at least to a large extent, only one iron is reduced. "To a large extent" means that, if there is an equilibrium between two species with one or the other iron reduced, this equilibrium would be far from 50/50. It is possible, however, that the difference in energy between the two reduced species, differing by the labeling of the iron ions is not orders of magnitude larger than kT. The equilibrium between these species must be fast on the NMR time scale. If it were slow, even small fractions of a second species would be detected by NMR. This is not the case. It is easy to simulate the effect, in the NMR spectra, of fast equilibrium between two species differing in the iron bearing the extra electron. This can be accomplished by introducing a difference in energy, Δ, between two energy level ladders, generated by Hamiltonian (1), which are equal except for the location of the reduced site. Fig. 5C (dashed line) shows the calculated temperature dependences obtained using the same A/h and J values as before, and $\Delta = 300$ cm^{-1} (corresponding to about 40 mV difference in reduction potentials for the two iron ions). It appears that the experimental behavior is better approximated, and, in particular, the antiCurie signals are now downfield. $\Delta = 300$ cm^{-1} results in a $\cong 20/80$ ratio between the two species at room temperature. AntiCurie downfield signals could also be obtained by considering only one species, and including the effect of an excited electronic quintet for iron(II), with a large difference in the J values for the two levels [12].

No NMR spectra are available for model compounds. Low temperature Mössbauer spectra on reduced samples obtained by reduction in situ of the $[Fe_2S_2(S_2\text{-}o\text{-}xyl)_2]^{2-}$ compound, containing bidentate thiolate ligands, show that the localized electronic configuration is an intrinsic property of $[Fe_2S_2(SR)_4]^{3-}$ clusters [63].

The next question in the characterization of reduced $[Fe_2S_2]$ proteins is that of knowing what is the reducible, or at least more easily reducible, iron. Through NOE and NOESY spectra on reduced $[Fe_2S_2]^{1+}$ proteins, the pairwise assignment of the β-CH$_2$ protons of the cysteines bound to the iron-sulfur core has been performed [219, 220].

Proteins containing the $[Fe_2S_2]$ center show high homology in their aminoacid sequences [221], and, in particular, the invariance of the four cysteine residues Cys 41, Cys 46, Cys 49, and Cys 79. Two β-CH$_2$ protons of Cys 41 and Cys 46, bound to the iron ion more exposed to the solvent, are so close (< 3 Å) that dipolar connectivities can be observed. The observation of NOE between two β-CH$_2$ protons whose signals have antiCurie behavior unambiguosly indicates that iron(II) is bound to Cys 41 and Cys 46 (Fig. 5D) [219, 220]. This finding could not be obtained using X-ray crystallography, because structure resolution does not allow us to discriminate between iron(III) and iron(II).

The following question is why that particular iron ion is reduced. The most noticeable difference between the two iron ions, in the structurally investigated proteins, is that the sulfur donors of Cys 41 and 46 form a larger number of hydrogen bonds than the cysteine sulfurs bound to the other, more buried, iron ion [13–16]. This is surely a reason for raising the reduction microscopic potential of the reducible iron. In addition, the effects of the distortions from the tetrahedral symmetry have been investigated through ab initio MO calculations [222]. These calculations indicate that, by starting from two iron(III) and adding one electron, the actual geometrical distortion gives rise to a larger Mulliken charge on the reducible iron, although selectivity is expected to be small. Selectivity increases if the electrostatic potential of each atom of the polypeptide chain is considered [222]. These effects easily account for the modest differences in reduction potential expected if the temperature dependence of the NMR signals is interpreted in terms of equilibrium between two species differing in energy by a few hundreds of wavenumbers (see above). If we follow the reasoning used by Warshel to explain the difference in oxidation state stabilization between HiPIPs and Fds [223], we can say that the effects of the permanent dipoles associated with CONH groups determines this difference in potential.

Some studies of the factors affecting the reduction potentials in these proteins, using site-directed mutagenesis, have started to appear in the literature [224]. Further comments on the factors determining the redox potentials will be given later.

The $[Fe_2S_2]^+$ system is quite exacting because it allows us to tackle the problem of B vs. J, electronic relaxation times, the sign and the absolute values of the hyperfine coupling at the individual iron sites, the factors determining the microscopic potentials, and to understand EPR. Again, significant progress has been achieved.

5 Oxidized Three Iron-Four Sulfur Ferredoxins $[Fe_3S_4]^+$

The next step in the differentiation of the iron ions is represented by the oxidized $[Fe_3S_4]^+$ systems. They formally contain three high spin iron(III) ions and the theoretical approach to describe magnetic coupling is that of Eq. (6). This spin coupling model requires the presence of three antiferromagnetic exchange coupling parameters almost equal in magnitude [225]. If the three J values necessary to describe the system are absolutely equal, then a double degeneracy of $S' = \frac{1}{2}$ levels is obtained. The requirement of equal J's cannot be fulfilled in proteins, because they provide an asymmetric environment. As soon as one J value becomes different, a conceptual barrier is overcome. Such a barrier is known as spin frustration. If we refer to Fig. 6 it soon appears that if spin 1 and spin 2 are antiferromagnetically coupled to spin 3, then spins 1 and 2 cannot be antiferromagnetically coupled to each other. It is enough that J_{12} is, for instance, only slightly smaller than J_{13} and J_{23} for the stronger antiferromagnetic couplings to force spins 1 and 2 to be ferromagnetically coupled (Fig. 6). On the other hand, if J_{12} is slightly larger than J_{13} and J_{23}, then S_1 and S_2 tend to be antiferromagnetically coupled, and S_3 cannot be either ferro- or antiferromagnetically coupled to both S_1 and S_2. The antiferromagnetic coupling does not provide a subspin $S_{12} = 0$ but a value just smaller than $\frac{5}{2}$, so that a total $S' = \frac{1}{2}$ is obtained.

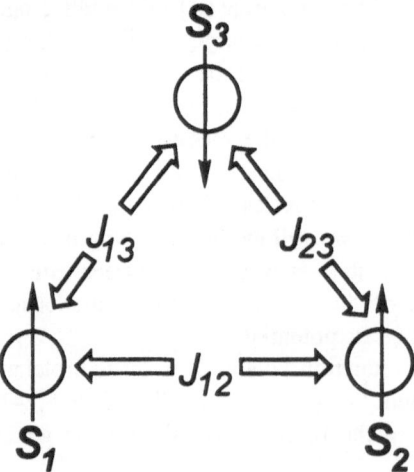

Fig. 6. Spin frustration in a trimetallic system

In order to have a simple theoretical tool, Hamiltonian (6) can be written in this case as [225]

$$H = J \sum_{i \neq j} \mathbf{S}_i \cdot \mathbf{S}_j + \Delta J_{12} \mathbf{S}_1 \cdot \mathbf{S}_2 \tag{15}$$

in such a way that the energies are provided by the following relationship:

$$E(S_{12}, S') = \tfrac{1}{2}[JS'(S' + 1) + \Delta J_{12} S_{12}(S_{12} + 1)] \tag{16}$$

where $J = J_{13} = J_{23}$ and $\Delta J_{12} = J_{12} - J$. It can easily be shown that the ground state is $S' = \tfrac{1}{2}$ for a large range of J values. In practice, $S' = \tfrac{1}{2}$ is the ground state in all known cases [41, 107, 226]. The EPR g values (Table 2) are quite isotropic, and are close to the value of the free electron. Note that again they are related to the monomeric values through equation [225]:

$$g = C_{11} g_1 + C_{21} g_2 + C_{31} g_3 \tag{17}$$

where C_{il} are analogous to those defined in Eq. (5), and g_i are the individual values of the uncoupled iron ions.

The isomer shifts observed in the Mössbauer spectra (Table 1) are indicative of Fe^{3+} oxidation states, and the magnetic Mössbauer data show that, in general, one iron ion has a very large negative, and one has a large positive, hyperfine coupling constant, whereas the third iron ion has a much smaller hyperfine constant, positive or negative [41, 104, 105, 107, 108]. These values have been interpreted as an indication that there is one high spin iron(III) ($S_3 = \tfrac{5}{2}$) antiferromagnetically coupled to a spin pair $S_{12} = 2$, in which one spin (for example S_2) is oriented antiparallel to S_3, and the other spin (S_1) is oriented almost perpendicular to the net magnetic moment [225].

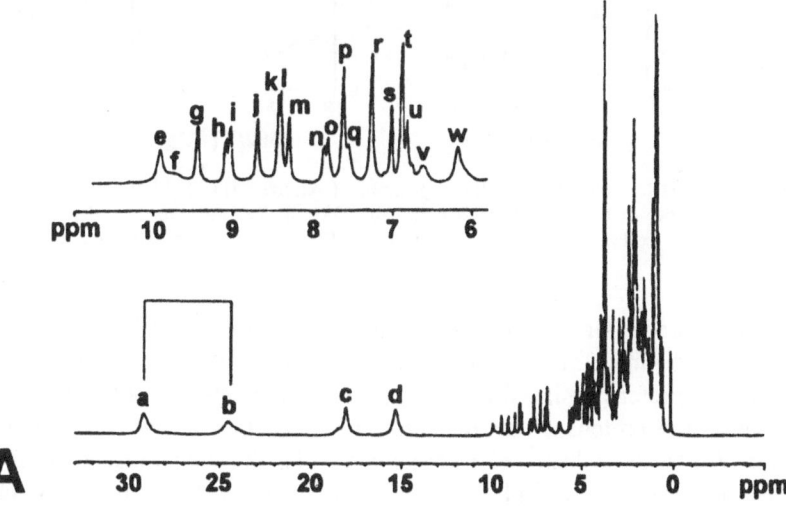

Fig. 7. continued

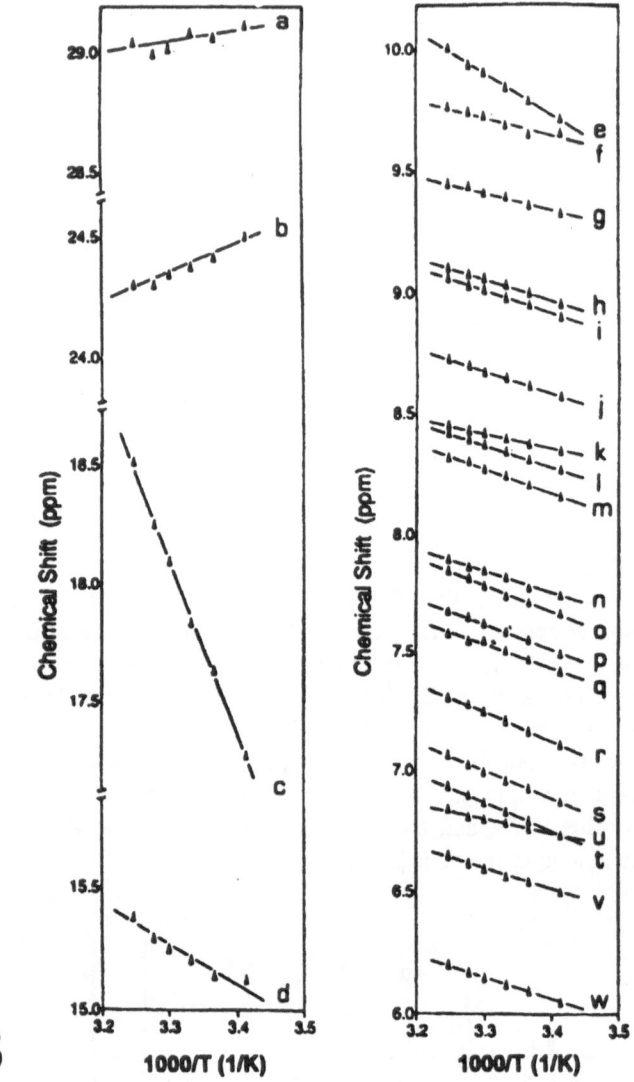

Fig. 7 A. 303 K, 300 MHz ^{1}H NMR spectrum of oxidized [Fe$_3$S$_4$]$^{1+}$ *D. gigas* Fd II. The geminal connectivity of Cys 50 β-CH$_2$ is indicated [228]; **B.** Temperature dependence of the ^{1}H NMR resonances of oxidized [Fe$_3$S$_4$]$^{1+}$ *D. gigas* Fd II. The resonances are labeled as in **A** [228]

Two recent NMR studies on proteins containing the [Fe$_3$S$_4$]$^{1+}$ core have revealed a distinct 2:1 magnetic asymmetry among the three iron ions [227, 228]. Only the protons assigned to one cysteine residue showed a Curie-type behavior, whereas the protons of the remaining two cysteines showed anti-Curie behavior (Fig. 7). In the scheme of Hamiltonian (15), this experimental result indicates that one iron ion, bound to the cysteine showing Curie behavior, has a subspin S_3 larger than the other subspin S_{12}. Calculations based on experimental

^{1}H NMR temperature dependences have produced the value of $J \cong 300 \text{ cm}^{-1}$, with $\Delta J_{12} = 6 \text{ cm}^{-1}$ [228]. The value of A/h for the uncoupled system was presumably taken equal to 1 MHz, although it is not stated in the paper. The difference in hyperfine coupling constant between Mössbauer and NMR is probably ascribed to the different operational temperature [143]. The value of $J = 40 \text{ cm}^{-1}$, proposed on the basis of the changes with temperature of the lineshape and intensity of the EPR signal [229, 230], is not compatible with magnetic susceptibility and magnetization studies carried out on oxidized *D. gigas* Fd II, which indicate that $J > 200 \text{ cm}^{-1}$ [226]. The NMR study on the proteins from *P. furiosus* and *T. litoralis* has shown that the two electronic distributions are different with respect to the frame of the protein backbone, although a complete sequence-specific assignment has not been performed [227]. On the other hand, in the case of *D. gigas* Fd II, a partial sequence-specific assignment has been tentatively carried out, which has indicated that the iron ion showing Curie behavior is bound to Cys 50 [228].

An $[Fe_3S_4]^+$ center has been obtained from *C. pasterianum* [231] and from *C. acidi urici* [232], and the relative NMR spectra are significantly different from the corresponding center in *D. gigas*.

In principle, two structures may occur, in which one J can be larger or smaller than the other two J's. The former case has already been described, and in this case four protons (two Cys β-CH$_2$) experience antiCurie behavior. The latter case would provide four protons experiencing Curie behavior.

6 Oxidized Four Iron-Four Sulfur HiPIP [Fe$_4$S$_4$]$^{3+}$

We will start the discussion of this class of proteins by considering the HiPIP II from *E. halophila* because it is the most symmetric, as will appear later. It shows a single axial EPR spectrum typical of $S' = \frac{1}{2}$ ground state with g values of 2.146 and 2.030 [124]. The Mössbauer data indicate two different sets of iron ions in a 1:1 ratio, with isomer shifts of 0.27 and 0.37 mm/s (Table 1) [124]. The former value is typical of a Fe(III) ion, for which typical values in this kind of environment range between 0.25 and 0.32 mm/s (Table 1). However, the 0.37 mm/s value is lower than that found for Fe(II) in rubredoxins (0.65–0.73 mm/s, Table 1) and in proteins containing the $[Fe_2S_2]^+$ center (0.48–0.56 mm/s, Table 1). It has been proposed that the value of 0.37 mm/s is associated with $Fe^{2.5+}$ mixed valence ions [112–114] (see also Sect. 3). It should also be noted that the difference in isomer shifts between Fe(III) and Fe(II) ions becomes smaller, the larger is the polymetallic center (Table 1). This is a consequence of the fact that a larger cluster provides a larger cage for charge delocalization. Indeed, ab initio MO calculations in the $[Fe_2S_2]$ centers have shown that, even though the extra charge added when passing from the oxidized to the reduced protein preferentially goes on the iron(II), it is largely distributed all over the $[Fe_2S_2]$ core

[222]. In other words, the definition based on oxidation states represents a rough approximation in terms of electronic charges. In general, the iron(III) and iron(II) nature of the ions is better characterized by the $S = \frac{5}{2}$ and $S = 2$ spin states, respectively. The concepts of charge and spin are well separated. This should be kept in mind in order to fully appreciate the present description of the electronic state of iron-sulfur proteins.

Magnetic Mössbauer provides two different sets of hyperfine coupling values between ^{57}Fe and the electron spin of the cluster [124]. The two average hyperfine values have opposite signs, as observed in the case of $[Fe_2S_2]^+$ centers [17]. The positive value is associated with the pair with smaller isomer shift (Fe^{3+}) and the negative value with the pair with larger isomer shift ($Fe^{2.5+}$). This means that the two $Fe^{2.5+}$ (subspin S_{34}) have larger subspins than the two Fe^{3+} (subspin S_{12}) [121]. The two subspins are antiferromagnetically coupled to provide $S' = \frac{1}{2}$. In the early years, subspin values $S_{12} = 4$ and $S_{34} = \frac{9}{2}$ were proposed [121], but the agreement between experimental and calculated ^{57}Fe hyperfine shifts was quantitatively unsatisfactory (see below).

The Hamiltonian of Eq. (7) [82, 142] allows us to handle any model with up to three different J values among the four iron ions (Fig. 8A). The energies are given by Eq. (8).

To account for a spin S_{34} larger than S_{12} it is necessary that J_{12} is larger than J and J_{34} is smaller than J, that is, $\Delta J_{12} > 0$ and $\Delta J_{34} < 0$. Under a large range of J values, $S_{34} = \frac{9}{2}$ and $S_{12} = 4$ are obtained. Again, in order to have equivalence between one iron(III) and one iron(II), electron delocalization is needed. In analogy with the treatment discussed in Sect. 3, a double exchange term can be included in the expression of the energies [167]:

$$E(S_{12}, S_{34}, S') = \tfrac{1}{2}[JS'(S' + 1) + \Delta J_{12}S_{12}(S_{12} + 1) + \Delta J_{34}S_{34}(S_{34} + 1)]$$

$$\pm B_{34}(S_{34} + \tfrac{1}{2}) \tag{18}$$

The parameter B_{34} is a ferromagnetic contribution, so that, when we choose to treat the system with Hamiltonian (7), which is without this latter term, the effect of B_{34} is similar to that of reducing the effective J_{34} value [142, 143]. It is easily shown from Eq. (8) and (18) that similar values are obtained with a larger B_{34} in Eq. (18) or smaller J_{34} in Eq. (8).

In oxidized HiPIPs spin frustration is operative. In fact, as shown in Fig. 8, if Fe$_1$ is strongly antiferromagnetically coupled to any other pairs, the latter ones result ferromagnetically coupled, even though their interaction is antiferromagnetic in nature. The situation is similar to that described for $[Fe_3S_4]^+$ systems (see Sect. 5). Under these circumstances, even a relatively small B induces electron delocalization over the J_{34} pair. The very first theoretical investigation relied on only two sets of J values ($\Delta J_{12} > 0$; $\Delta J_{34} = 0$) and made use of a very large value of B, of the order of 500 cm^{-1} [165, 166]. Whatever the approach, once we have the ground state we can define an expectation value of the total S'_z, $\langle S'_z \rangle$, and an expectation value of the S_i for each metal site, $\langle S_{iz} \rangle$. In analogy with

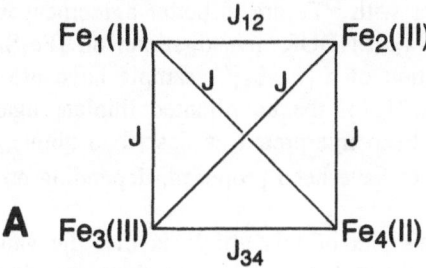

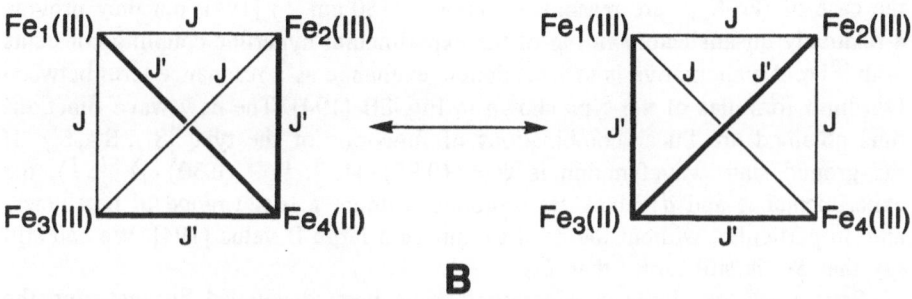

Fig. 8 A. Arrangement of the ferric and ferrous ions, and of their Heisenberg exchange parameters in oxidized HiPIP under C_{2v} symmetry [194]; **B.** Illustration of the resonance between two limit formulas characterized by a C_{3v} symmetry (see text) [194].

Table 3. Experimental and calculated[a] hyperfine constants and g_{av} values for oxidized *E. halophila* HiPIP II [194]

	Experimental values	C_{2v} $|4, \frac{9}{2}, 1/2\rangle$	C_{2v} $|3, \frac{7}{2}, 1/2\rangle$	C_{3v}	$C_{3v} \rightarrow C_s^b$
A_{12}(MHz)	21.4 ± 1.5	26.7	20.0	−3.1	21.5
A_{34}^c(MHz)	-31.5 ± 1.0	−38.3	−31.5	8.9	−33.1
g_{av}	2.07	2.054	2.042	2.065	2.066

[a] All calculations performed using monomer hyperfine constants $a(Fe^{3+}) = -20$ MHz and $a(Fe^{2+})$ −22 MHz [89]
[b] Calculated using $J' = \frac{1}{2}J$, $\Delta J_{12} = \Delta J_{34} = -0.1J$
[c] Average value of A_3 and A_4

Eq. 5, the ratios $\langle S_{iz}\rangle/\langle S_z'\rangle$ tell us how much the hyperfine coupling is changed upon going from an isolated iron ion to the magnetically coupled system. It is thus possible to calculate the hyperfine coupling in the polymetallic center using the A value of rubredoxins for the uncoupled system, and then to test whether the calculated spin wave function is adequate. The result for various $|S_{12}, S_{34}, S'\rangle$ is shown in Table 3. A large value of B (e.g. 500 cm^{-1}) renders then also possible an $S_{34} = \frac{7}{2}$ and $S_{12} = 3$ ground state [194]. Unfortunately there is no independent estimate of B. With this new ground state spin wave function, the calculated hyperfine coupling values with ^{57}Fe are in better agreement with the experimental values (Table 3) [194]. ENDOR investigations on $[Fe_4S_4]^{3+}$ model centers produced upon γ-irradiation of a $[Fe_4S_4]^{2+}$ sample have provided the hyperfine values between ^{57}Fe, or ^{1}H of the coordinated thiolate ligands, and the cluster spin. The data have been interpreted as described above, and both $|\frac{9}{2}, 4, \frac{1}{2}\rangle$ and $|\frac{7}{2}, 3, \frac{1}{2}\rangle$ ground states have been proposed, depending on the particular case [87, 233–235].

The feeling of the authors of this article is that the value of B cannot be much different from a polymetallic center to another, even if the number of ions changes. In other words, only B values of the order of the upper limit estimated in the case of $[Fe_2S_2]^+$ are reasonable (approx. 150 cm^{-1}) [194], but they provide a relatively unsatisfactory fitting of the experimental hyperfine coupling constants with ^{57}Fe. An alternative is to treat double exchange as a resonance term between two limit formulas of the type shown in Fig. 8B [194]. The new wave functions thus obtained are linear combinations of functions of the type $|S_{12}, S_{34}, S'\rangle$. If the ground state wavefunction is $\psi = (0.95) \cdot |4, \frac{9}{2}, \frac{1}{2}\rangle - (0.30) \cdot |4, \frac{7}{2}, \frac{1}{2}\rangle$, the experimental A and g values are reproduced under a broad range of parameters and, in particular, without the need to impose a large B value [194]. We can still say that S_{34} is still larger than S_{12}.

Several of the above considerations have been developed by analyzing the ^{1}H NMR spectra. The ^{1}H NMR spectrum of oxidized E. halophila HiPIP II is shown in Fig. 9A [139, 236]. NOEs have allowed to establish that A–C,

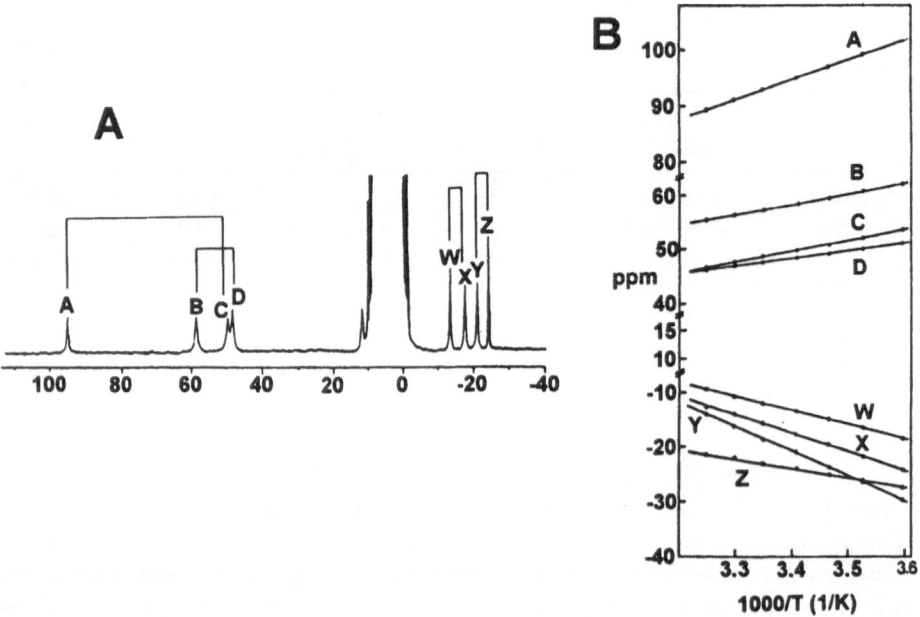

Fig. 9 A. 298 K, 600 MHz ^{1}H NMR spectrum of oxidized $[Fe_4S_4]^{3+}$ HiPIP II from E. halophila. The cysteine β-CH$_2$ geminal connectivities are indicated [143]; **B.** Temperature dependence of the ^{1}H NMR resonances of oxidized HiPIP II from E. halophila. The resonances are labeled as in **A** [143].

B–D, W–X and Y–Z are the signals of the geminal cysteine β-CH_2 bound to the cluster [236]. Therefore, two cysteines experience downfield shifts and negative hyperfine coupling, whereas the other two cysteines upfield shifts and positive hyperfine coupling. By following the interpretation of magnetic Mössbauer and in analogy with the iron(III)-iron(II) case, we assign the upfield shifted signals to cysteines bound to iron(III) with S_{12} subspin smaller than the S_{34} of the mixed valence pair. The nuclei of the latter domain experience downfield shifts. The temperature dependence of the shifts, shown in Fig. 9B, has been qualitatively reproduced with $J = 300$ cm^{-1}, $J_{12} = 400$ cm^{-1}, and $J_{34} = 200$ cm^{-1} [236].

An equally satisfactory fitting for the A values is obtained when using four J values [194]. It is worth remembering here that delocalization on only a pair of iron ions rather than on all cluster ions is an experimental result, even though, within the approximation of the present theory, delocalization on the other pair to an extent smaller than $\cong 10\%$ cannot be excluded. Delocalization on a given pair, instead of on other pairs, is apparently the result of either a fine geometrical distortion of the core or of the electrostatic potential generated by the atoms of the polypeptide chain, or both. Eventually, this has to be related to the overall reduction potential.

Low symmetry oxidized HiPIP $[Fe_4S_4]^{3+}$

Besides the HiPIP II from *E. halophila*, the other HiPIPs investigated by NMR, i.e. HiPIP I from *E. halophila* [139], HiPIP I and II from *E. vacuolata* [139, 175], HiPIPs from *C. vinosum* [138, 142, 146, 173, 177], *R. gelatinosus* [138, 143, 144], *R. tenue* [140], *C. gracile* [141], *R. globiformis* [147], and *Rf. fermentans* [151] are characterized by asymmetric hyperfine shifts of the β-CH_2 protons. The ^{1}H NMR spectra of some low symmetry HiPIPs (Fig. 10B–E) show two β-CH_2 signals downfield and two upfield at values similar to those observed for the symmetric HiPIP II from *E. halophila* (Fig. 10A), together with four β-CH_2 downfield, not too far shifted. Actually, in all cases the least downfield pair has an antiCurie temperature dependence of the type observed in reduced $[Fe_2S_2]^{1+}$ proteins (Sect. 4). NOE measurements have established the pairwise nature of the signals, respectively two upfield, two downfield, two downfield with antiCurie behavior, and two more downfield but not shifted much [142, 143, 147, 151, 175, 177]. The first interpretation of these spectra was that, as in the case of HiPIP II from *E. halophila*, there is a mixed valence pair with downfield shifts, and one iron(III) ion with upfield shifts of the β-CH_2 protons. The pair of downfield shifted signals with antiCurie behavior would be due to the β-CH_2 of a cysteine bound to an iron largely iron(III) but somehow involved in a three center electron delocalization [142, 236, 237]. Mössbauer spectra have been shown to be insensitive to this delocalization. This picture is represented in Fig. 11A.

The sequence specific, stereospecific assignment of the protons of the β-CH_2 of the cysteines bound to the iron ions has been obtained for all the above oxidized HiPIPs [144, 146, 147, 151, 173, 175]. It is worth mentioning to mention

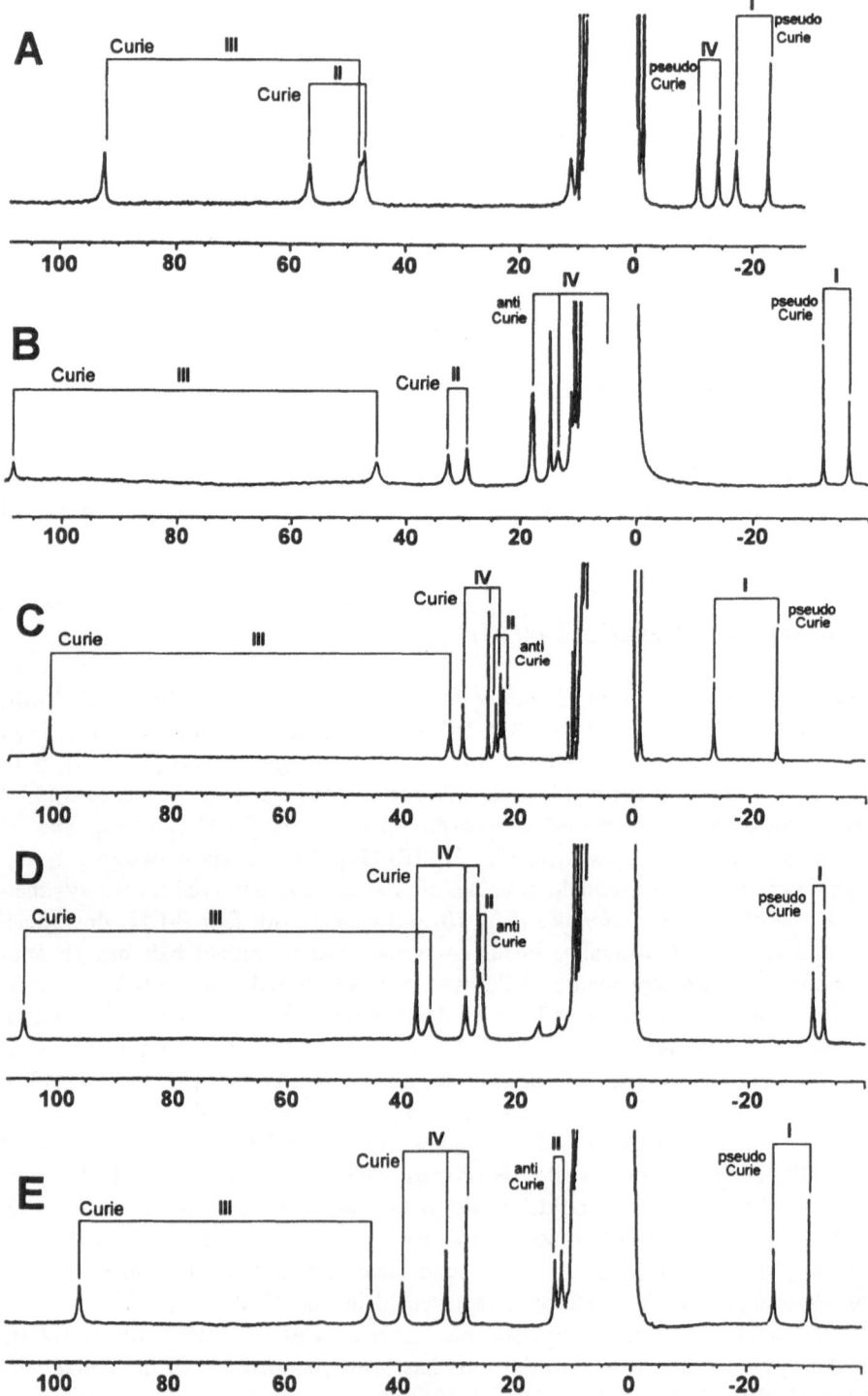

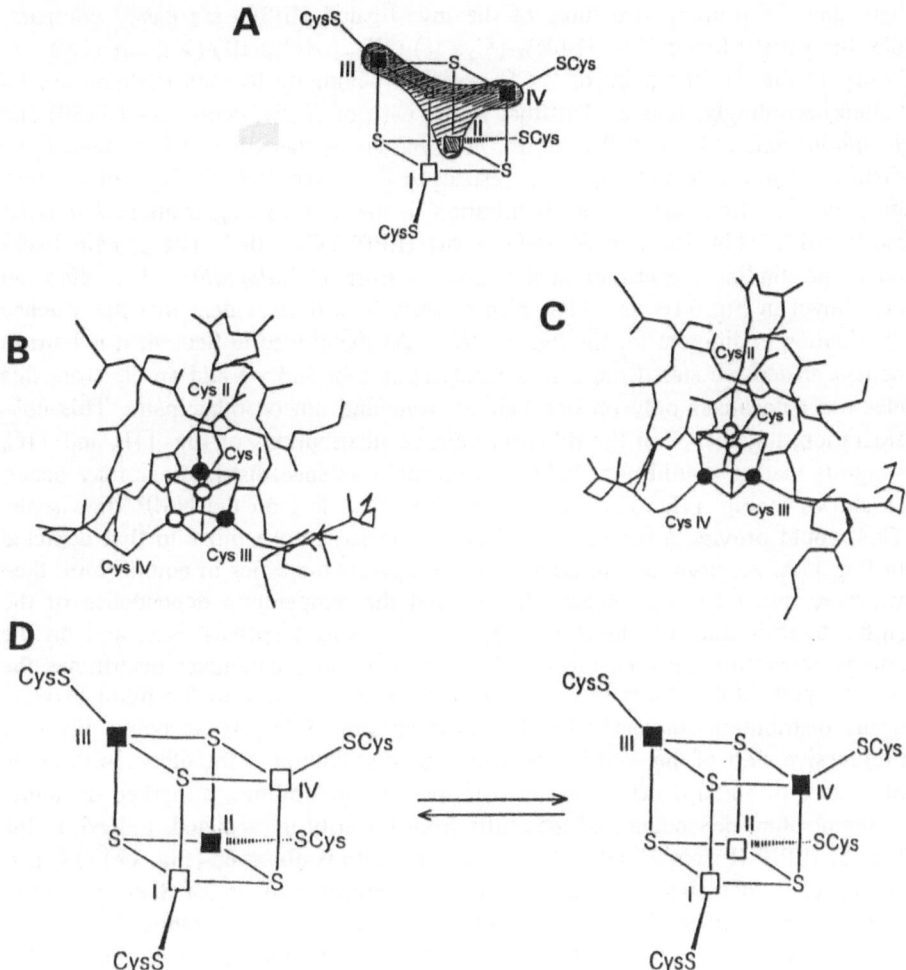

Fig. 11 A. Possible low symmetry situation in the oxidized $[Fe_4S_4]^{3+}$ cluster in HiPIPs. This situation could be representative, for example, of the HiPIP from *C. vinosum*. The geometric arrangement of the oxidized $[Fe_4S_4]^{3+}$ cluster and the surrounding residues in *E. halophila* HiPIP II derived from molecular dynamics calculations (**B**) [173], and *C. vinosum* HiPIP derived from crystallographic data (**C**) [22, 23, 25] are shown. *Empty circles* indicate the ferric pair, whereas *filled circles* the mixed valence pair, as derived from NMR spectroscopy [147, 175]; **D.** Possible low symmetry situation in the oxidized $[Fe_4S_4]^{3+}$ cluster in HiPIPs as obtained in the presence of chemical equilibrium between two situations of higher symmetry, of the type of **B** and **C**. *Roman numerals* refer to cysteine numbering in the sequence (see text)

Fig. 10 A–E. 300 K, 600 MHz ^1H NMR spectra of oxidized $[Fe_4S_4]^{3+}$ HiPIP from; **A.** *E. halophila* (iso-II) [139, 236]; **B.** *R. globiformis* [147]; **C.** *E. vacuolata* (iso-II) [139, 175]; **D.** *C. vinosum* [142, 162]; **E.** *R. gelatinosus* [143, 144]. The cysteine β-CH$_2$ geminal connectivities are indicated and labeled according to the primary sequence [238] (see text). Cysteine IV is a triplet because the α-CH is also assigned. The temperature dependence of each pair is also indicated

here that the primary structures of the investigated HiPIPs are easily comparable, the pattern being (Cys I)-$(X)_2$-(Cys II)-$(X)_{9-16}$-(Cys III)-$(X)_{13-15}$-(Cys IV)
[238]. In Fig. 10, the pairs of β-CH_2 protons belonging to each cysteine are labelled accordingly. It is evident that in the case of *E. halophila* iso-II [159] and
R. globiformis [147] HiPIPs, the iron ions oxidation states have been found to be
distributed in a different way with respect to the other three HiPIPs, with a progressive shift from the former distribution to the latter, going from *E. halophila*
iso-II HiPIP (Fig. 10A) to *R. gelatinosus* HiPIP (Fig. 10E). The protein backbone surrounding the cluster in the proteins from *E. halophila* and *C. vinosum*
are shown in Fig. 11B and 11C, respectively. It is thus evident that the valence
distribution is different for the two proteins. As mentioned in Sect. 4, it is hard to
believe that if we start from a four iron(III) system and we add an electron, this
electron delocalizes only on one pair of irons and not on other pairs. This consideration, together with the different valence distributions of Fig. 11B and 11C,
suggests that an equilibrium between two main valence distributions may occur,
as shown in Fig. 11D. The equilibrium would be fast on the NMR time scale.
This would provide a time-averaged electronic structure similar to that depicted
in Fig. 11A. As discussed in Sect. 3, if we have two species in equilibrium, then
we have two ladders of energy levels, and the temperature dependence of the
shifts depends both on the J values, as in the single species case, and on the
energy separation between the two ladders. The latter parameter determines the
relative population of the two species. This is consistent with the trend of electronic distributions indicated by the NMR spectra of Fig. 10 as being due to a
progressive shift of the equilibrium from one distribution to the other. In the case
of a close to 50/50 distribution between the two possibilities, a marked deviation
of temperature dependence of the shifts from linearity is predicted. Indeed, in the
case of HiPIP II from *E. vacuolata*, such curvature is observed (Fig. 12) [175]. A
similar effect has also been observed upon substitution of Se for S in the cluster
of *C. vinosum* HiPIP, for which a slight curvature of the temperature dependence
has indicated that such equilibrium was influenced by the Se substitution [239].

    We can plot the shifts of the pair with antiCurie behavior vs. the distribution
of species (Fig. 13) [176]. This plot can be qualitatively used to decide the
distribution of spins in other HiPIPs.

    Some available EPR spectra of HiPIPs are shown in Fig. 14. The striking
feature of these spectra is that, as opposed to the spectrum of *E. halophila*
HiPIP II (Fig. 14A) [124], the spectra of *C. vinosum* (Fig. 14B) [206, 207],
E. vacuolata iso-II (Fig. 14C) [175], and *R. gelatinosus* (Fig. 14D) [208] HiPIP
show several additional features whose origin is still uncertain. The EPR spectrum
of *C. vinosum* HiPIP (Fig. 14B) was first interpreted as due to two components
of equal intensity, one axial species with $g_{\parallel} = 2.120$ and $g_{\perp} = 2.040$, and one
rhombic species with $g_1 = 2.086$, $g_2 = 2.055$, and $g_3 = 2.040$ [206]. More recently, this spectrum has been reinterpreted as due to one major species with
$g_1 = 2.115$, $g_2 = 2.037$, and $g_3 = 2.025$, one minor species with $g_1 = 2.129$, $g_2 =
2.071$, and $g_3 = 2.041$, together with some features at $g_1 = 2.114$, $g_2 = 2.038$,
and $g_3 = 2.024$ attributed tentatively to a dimerization of the protein [207].

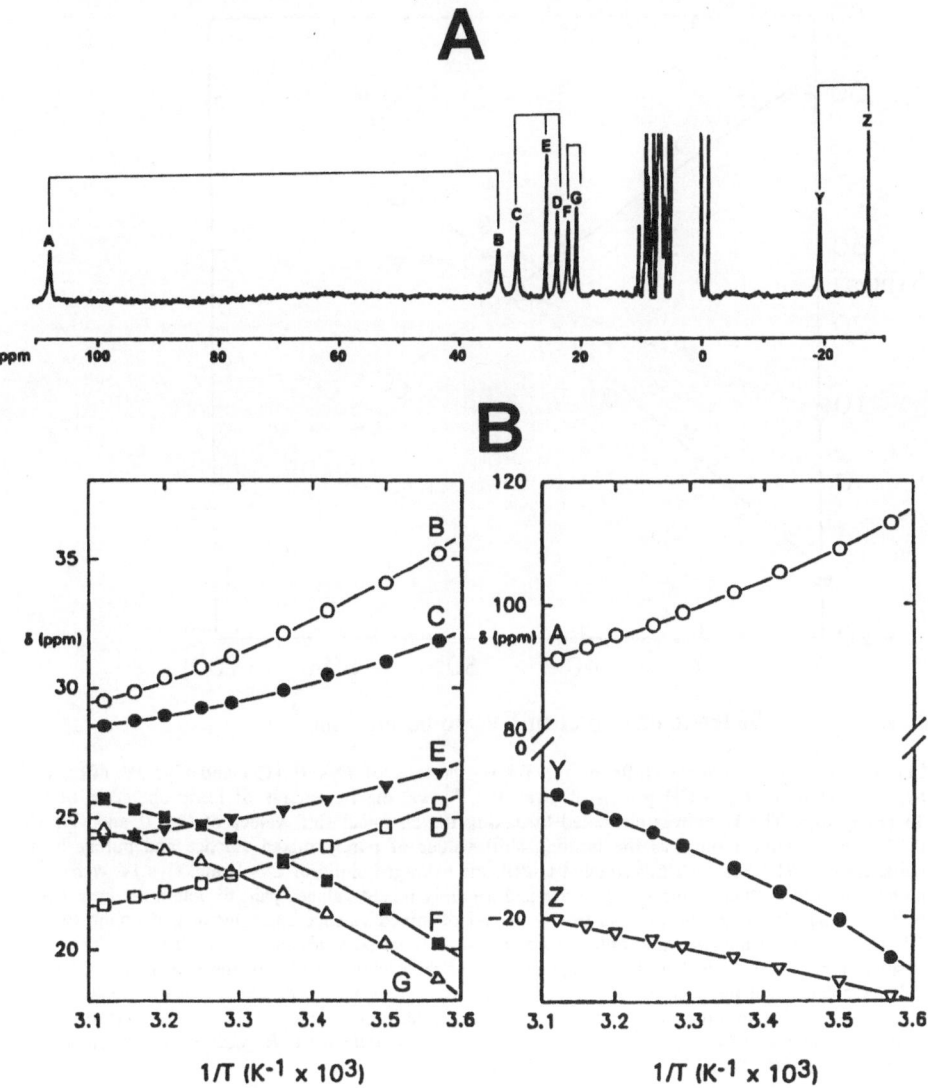

Fig. 12 A. 300 K, 600 MHz ^{1}H NMR spectrum of oxidized $[Fe_4S_4]^{3+}$ HiPIP II from *E. vacuolata*. The cysteine β-CH$_2$ geminal connectivities are indicated [175]; **B.** Temperature dependence of the chemical shifts of the hyperfine shifted signals in oxidized *E. vacuolata* HiPIP II. The signals are labeled as in **A** [175]

In contrast to the above differences among high and low symmetry HiPIPs, the 4.2 K Mössbauer spectrum of the *C. vinosum* protein [118, 121, 122] is virtually identical to that of the *E. halophila* HiPIP II [124]. This indicates that, whichever the reason for the differences observed in the NMR and EPR spectra of low and high symmetry HiPIPs, Mössbauer spectra do not have the resolution to distinguish between two or more species.

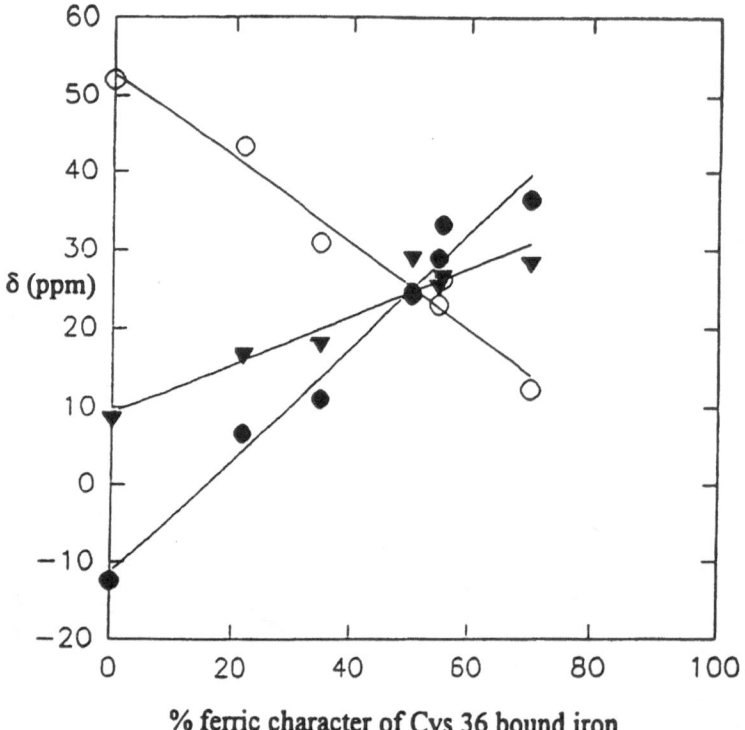

Fig. 13. Averaged chemical shifts of the β-CH$_2$ protons of Cys II (O) and Cys IV (●), and hyperfine shifts of the α-CH proton of Cys IV (▼), vs. the percentage of ferric character of cys 36 bound iron. The latter was estimated by taking the chemical shift values of Cys II and Cys IV β-CH$_2$ in the same protein as the limiting shift values of purely mixed valence and purely ferric pairs, respectively. For each investigated HiPIP, the averaged shifts of Cys II and Cys IV were then compared to the values of the purely ferric and a purely mixed valence pair, to obtain the percentage of ferric character of the iron bound to Cys II and of mixed valence character of that bound to Cys IV. The average of the two values obtained as above is reported in abscissa for each HiPIP. Thus, the abscissa reports the percentage ferric character of the iron bound to Cys II, the percentage of mixed valence character of the iron bound to Cys IV, and, by extension, the percentage of the cluster in the electronic configuration depicted on the right of the equilibrium in Figure 11D. The data (from left to right) refer to the HiPIPs from *E. halophila* iso-II, *E. halophila* iso-I, *R. globiformis*, *E. vacuolata* iso-I, *E. vacuolata* iso-II, *C. vinosum* and *R. gelatinosus*

The recent work on γ-irradiated [Fe$_4$S$_4$] models has shown that each center has a different EPR spectrum depending on the pair experiencing mixed valence character [87]. This would imply that the spectra of all HiPIP, except for HiPIP iso-II from *E. halophila*, are due to an overlap of spectra related to species differing in valence distribution.

Finally, the intense and very temperature dependent MCD spectra of oxidized HiPIP from *C. vinosum* have been utilized to confirm the $S' = \frac{1}{2}$ electronic ground state in proteins containing the [Fe$_4$S$_4$]$^{3+}$ core [240].

In summary, the only case so far described of an oxidized HiPIP containing a single species has been thoroughly analyzed and essentially understood on the

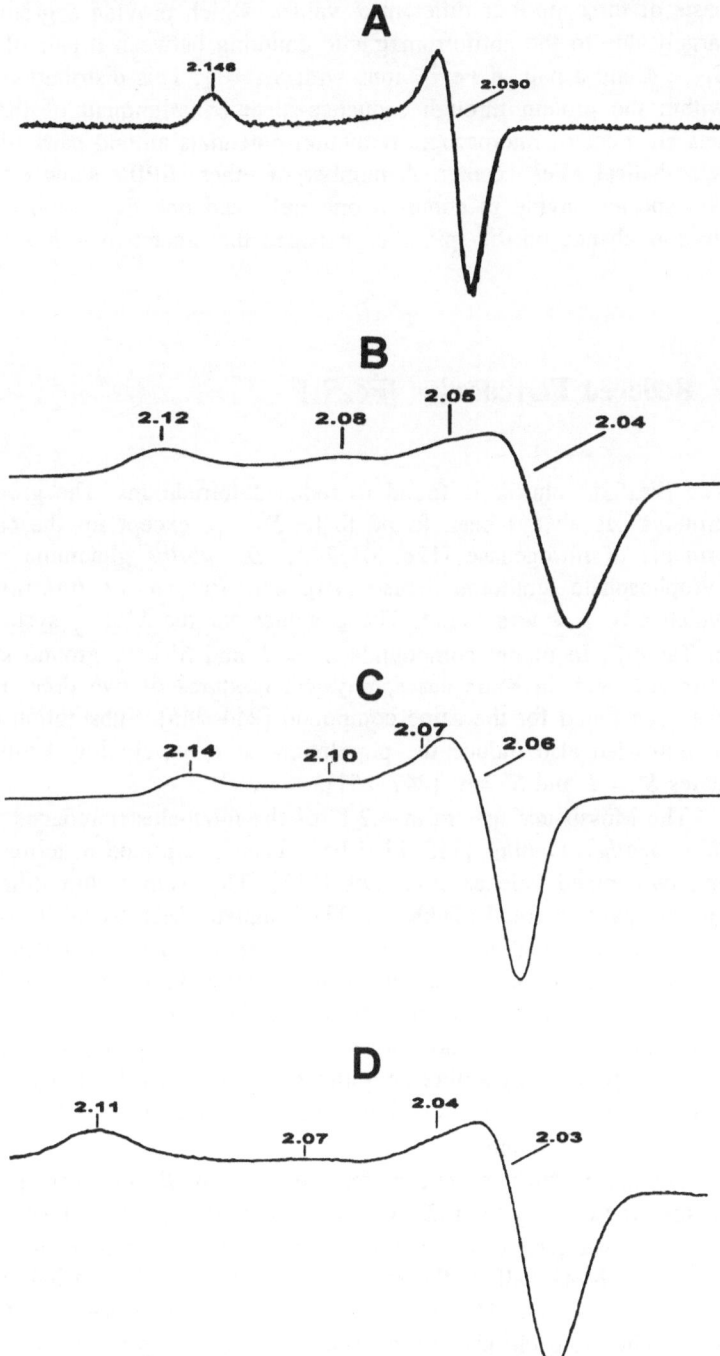

Fig. 14 A–D. EPR spectra of oxidized $[Fe_4S_4]^{3+}$ HiPIP from **A.** *E. halophila* (iso-II) [124, 208]; **B.** *C. vinosum* [206, 207]; **C.** *E. vacuolata* (iso-II) [175]; **D.** *R. gelatinosus* [208]

basis of three or four different J values, which provide a ground $S' = \frac{1}{2}$ state largely due to the antiferromagnetic coupling between a pair of Fe^{3+} ions with $S_{12} = 4$ and a pair of $Fe^{2.5+}$ ions with $S_{34} = \frac{9}{2}$. This distribution of spins, found within the protein through sequence-specific assignment of the NMR signals, sets an order of microscopic reduction potentials among pairs of Fe^{3+} ions of a hypothetical $4Fe^{3+}$ center. A number of other HiPIPs show equilibria between two species having in common one Fe^{3+} and one $Fe^{2.5+}$ ion, and differing for the interchange on the spin state between the other two iron atoms.

7 Reduced Ferredoxins [Fe$_4$S$_4$]$^+$

The [Fe$_4$S$_4$]$^+$ cluster is found in reduced ferredoxins. The ground state in the proteins has always been found to be $S' = \frac{1}{2}$ except for the cases of the iron proteins of nitrogenase [126, 241, 242], *B. subtilis* glutamine phospho-ribosyl-pyrophosphate amidotransferase [10], and *Pyrococcus furiosus* Fd [243], for which a $S' \geqslant \frac{3}{2}$ was found. The g values for the $S' = \frac{1}{2}$ systems are reported in Table 2. In model compounds $S' = \frac{1}{2}$ and $S' = \frac{3}{2}$ ground states have been observed, and, in some cases, physical mixtures of two different ground states has been found for the same compound [244–246]. Substitution of S with Se in proteins can also induce the population, at relatively low temperature, of spin states $S' = \frac{3}{2}$ and $S' = \frac{7}{2}$ [247–251].

The Mössbauer spectra at 4.2 K of the monocluster reduced ferredoxin from *B. stearothermophilus* [113–115] have been interpreted in terms of two iron(II) and two mixed valence iron ions [113]. The isomer shift differences between the two pairs is small (Table 1). The magnetic Mössbauer clearly indicates two sets of inequivalent ions [113–115]. Mössbauer data are available also for the closely related *C. pasteurianum* dicluster ferredoxin. At zero field it is not possible to discriminate between a ferrous and a mixed valence pair; yet, magnetic Mössbauer shows the characteristic splitting in two pairs, one with negative and one with positive hyperfine coupling constants [116–117]. This splitting can be again rationalized in terms of two subspins, e.g. $\frac{9}{2}$ and 4, whose difference yields the total $S' = \frac{1}{2}$ value.

The proton NMR spectra of ferredoxins from *B. polymyxa* [252], *B. thermoproteolyticus* [253], and *B. stearothermophilus* [253], all containing a single [Fe$_4$S$_4$] cluster, show all downfield shifts, and the temperature dependence of the shifts shows half of the signals with slight Curie and half with slight anti-Curie behavior (Fig. 15A–C). In the case of *C. acidi urici* and *C. pasteurianum* ferredoxins, two closely spaced [Fe$_4$S$_4$] cores are present, and the temperature dependence of the ^{1}H NMR shifts (Fig. 15D–G) is such that half of the signals for each cluster follow a slight Curie behavior whereas half show a slight antiCurie behavior [154, 157, 158]. In the case of *C. acidi urici* ferredoxin, the

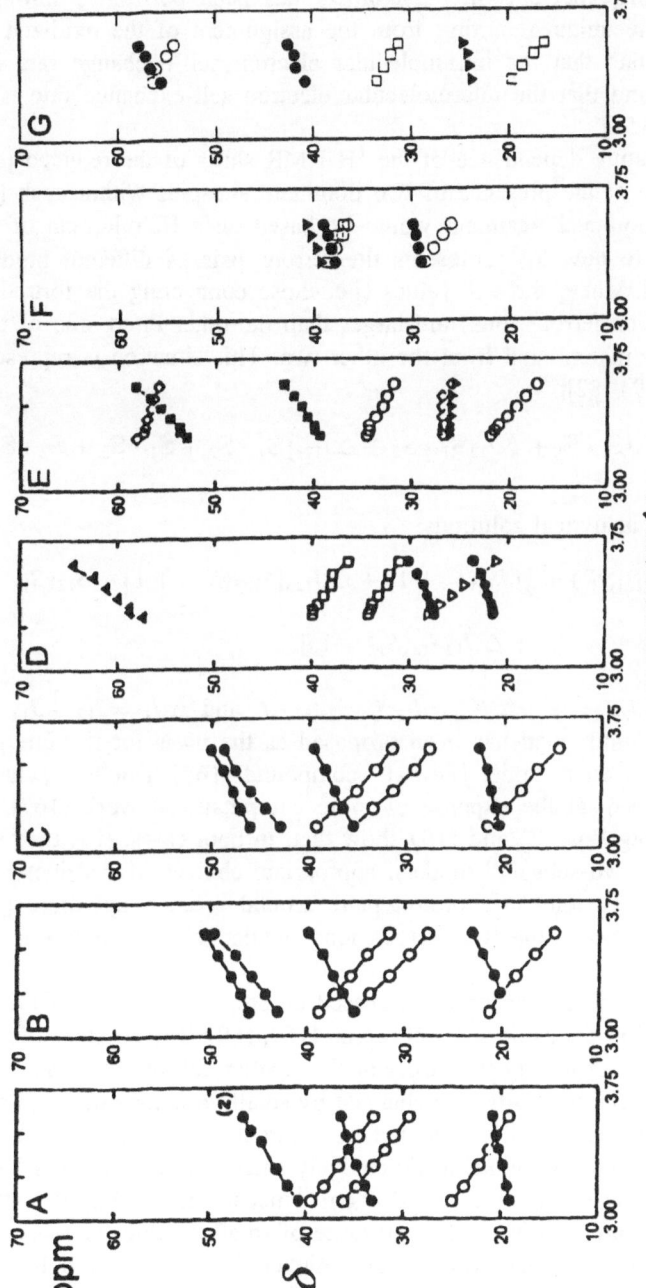

Fig. 15 A–G. Temperature dependences of the ^1H NMR hyperfine shifted signals of reduced $[Fe_4S_4]^{1+}$ ferredoxin from **A.** *B. polymyxa* (single cluster) [252]; **B.** *B. thermoproteolyticus* (single cluster) [253]; **C.** *B. stearothermophilus* (single cluster) [253]; **D.** *C. pasteurianum* cluster II [157]; **E.** *C. pasteurianum* cluster I [157]; **F.** *C. acidi urici* cluster II [158]; **G.** *C. acidi urici* cluster I [158]. *Hollow symbols* indicate antiCurie temperature dependence, whereas *filled symbols* indicate Curie behavior. In **D**, **E**, **F**, and **G**, identical symbols are used for resonances belonging to homologous cysteine β-CH_2 pairs in the two clusters.

stereospecific, sequence-specific, assignment has been performed through saturation transfer techniques starting from the assignment of the oxidized species, exploiting the fact that the intramolecular electron self-exchange rate is larger than 10^5 s^{-1}, and that the intermolecular electron self-exchange rate is smaller than 10^4 s^{-1} [158].

The temperature dependence of the ^{1}H NMR shifts of the reduced protein is again indicative of the presence of two dominant subspins within each $[Fe_4S_4]^+$ cluster. The theoretical treatment would be based on a Hamiltonian of the type of Eq. (7) where now S_{12} represents the ferrous pair. A different model could also be applied, where three J values (i.e. those connecting the formally ferric ion with the three ferrous ions) are larger than the other three; one of the latter can be further differentiated from the other two. This situation is represented by Hamiltonian (19) [82]:

$$H = \sum_{i\neq j} J\mathbf{S}_i \cdot \mathbf{S}_j + \Delta J_{12}\mathbf{S}_1 \cdot \mathbf{S}_2 + \Delta J_{123}(\mathbf{S}_1 \cdot \mathbf{S}_2 + \mathbf{S}_1 \cdot \mathbf{S}_3 + \mathbf{S}_2 \cdot \mathbf{S}_3) \quad (19)$$

which also has analytical solutions:

$$E(S_{12}, S_{123}, S') = \tfrac{1}{2}[JS'(S' + 1) + \Delta J_{123}[S_{123}(S_{123} + 1) - S_{12}(S_{12} + 1)]$$

$$+ \Delta J_{12}S_{12}(S_{12} + 1)] \quad (20)$$

where $J = J_{14} = J_{24} = J_{34}$, $\Delta J_{123} = J_{13} (=J_{23}) - J$, and $\Delta J_{12} = J_{12} - J_{13} (=J_{23})$.

The latter Hamiltonian has been proposed as the basis for the interpretation of magnetic data on a model $[Fe_4S_4]^+$ compound [167]. Double exchange can be also introduced, at the expense of more computational work [167]. Sample calculations using Eqs. (7) and (19) show that, in both cases, (i.e. the "trigonal" model and the "two-subspin" model), appropriate choices of J values can yield two Curie and two antiCurie proton pairs around room temperature [158]. In particular, a J between the two ferrous ions smaller than the others is required to fit the data.

At variance with all the cases discussed before, Curie and antiCurie signals are intermingled in the same spectral region [157, 158]. The energies of the states with the ferrous pair having the larger or the smaller subspin are very close, and the ground state can be easily interchanged by small variations in the parameters.

Although the whole picture is not as clearcut as in the case of oxidized HiPIPs, some features are evident: i) relatively small J values (around 100 cm^{-1}, as opposed to J values around 200–300 cm^{-1} needed to fit clusters with higher oxidation states) are necessary. The presence of small J values is consistent with the many observations that ground states higher than $S' > \tfrac{1}{2}$ can be obtained. The energies of the states with the ferrous pair having the larger or the smaller subspin are very close, and the ground state can be easily interchanged by small variations in the parameters. Of course, the detailed understanding of the experimental behavior may need the existence of equilibria among species with different valence distributions.

8 The $[Fe_3S_4]^0$ Core

This polymetallic center formally contains two Fe^{3+} and one Fe^{2+} ion. The ground state is $S' = 2$ [109, 226, 254]. Historically it is the first core for which the double exchange term was invoked [102, 169]. Mössbauer spectroscopy indicates that there is one iron(III) and two irons at the oxidation state 2.5+ (Table 1) [41, 103–106, 109]. The latter pair appears to be antiferromagnetically coupled to the ferric ion as the result of spin frustration, as explained in Sect. 5. In the literature there are two approaches which provide the $S' = 2$ ground state, one with equal J values and one considerable B value, and uses the Hamiltonian (21) [103]:

$$H = J[(^1S_1 \cdot {}^1S_2) + (^1S_1 + {}^1S_2) \cdot S_3] \cdot O_1 + J[(^2S_1 \cdot {}^2S_2)$$
$$+ (^2S_1 + {}^2S_2) \cdot S_3] \cdot O_2 + B_{12}V_{12}T_{12} \tag{21}$$

where the labeling is that reported in Fig. 6. The other approach affords similar results by using the Hamiltonian (15) with $\Delta J_{12} < 0$ and $S_1 = \frac{5}{2}$, $S_2 = 2$, and $S_3 = \frac{5}{2}$. The smaller value of J is that relative to the mixed valence pair for which a small, though arbitrary, B_{12} value is needed to have electron delocalization [255].

^{1}H NMR spectra are not available in the literature for these systems, probably because of the relatively large S value, which provides a larger broadening of the signals of protons bound to carbon atoms in α with respect to the donor atoms.

9 A Comment on the Heisenberg Approach

As outlined throughout this article, we have relied on the Heisenberg approach to account for the values of the ^{1}H NMR hyperfine shifts, their temperature dependence, and implicitly for the magnetic susceptibility data. Other hyperfine coupling energies obtained from ENDOR and magnetic Mössbauer are accounted for as well. We have also underlined that there is covariance between J values and the double exchange term B, so that the results obtained by using only J values can also be obtained by changing one J value and compensating for it with an appropriate B value. We would now like to make a comparative analysis of the J values needed to account for the experimental NMR data, in order to show that the overall interpretation is internally consistent. Throughout the present treatment, we have assumed that the hyperfine coupling constants for $S = \frac{5}{2}$ and $S = 2$ ions are the same; we estimate that variations, if present, do not exceed a factor two. Variations may also occur with the cluster core oxidation level. Furthermore, only one $S = 2$ state has been taken into consideration for iron(II).

The $[Fe_2S_2]^{2+}$ system has been treated with a J of 290 cm^{-1} (see Sect. 2). The hyperfine shifts are expected to be around 30–40 ppm, and to have an anti-Curie behaviour as shown in Figure 5C.

For the reduced $[Fe_2S_2]^+$ system a J value of 200 cm^{-1} has been used (see Sect. 4) to predict a shift for iron(III) of ca. 150 ppm with Curie behavior, and a shift of about 30 ppm upfield for iron(II). The intercept of both shifts at infinite temperature is sizably downfield. There are several ways to allow the β-CH$_2$ protons of the Fe^{2+} domain to go downfield and have an antiCurie behavior (see Sect. 4).

$[Fe_3S_4]^+$ systems have been accounted for by J values of the order of 300 cm^{-1} (see Sect. 5). A smaller value is needed in the case of $[Fe_3S_4]^0$ clusters, that mimicks the effect of spin delocalization in the mixed valence pair (see Sect. 7). Larger hyperfine shifts, both downfield and upfield, are predicted for this case.

In the case of $[Fe_4S_4]^{2+}$ core, an average J value of 150 cm^{-1} has been used. Two J values, corresponding to the two mixed valence pairs (see Sect. 3), had to be substantially lowered with respect to the other four. This yields an overall antiCurie behavior.

In the case of $[Fe_4S_4]^{3+}$ core an average J value of 300 cm^{-1} has been used previously (see Sect. 6) However, even sensibly reduced J values may account for the experimental behavior. In this case, one J value has been increased over the average, and the other decreased by the same amount. The pair with larger J is the ferric pair, and the other the mixed valence pair (see Sect. 6). The shifts of the β-CH$_2$ protons of the latter domain are sizably downfield and the other upfield.

In $[Fe_4S_4]^+$ systems J values around 100 cm^{-1} seem to be needed, consistent with the overall increase of the ionic radii (see Sect. 8). The main feature of the system is that the J value between the two ferrous ions has to be appreciably smaller than all the others. Such small J values account for low-lying excited states and therefore for all the shifts being downfield and none upfield. They may also account for the observation of higher spin ground states in some proteins, model compounds, and in selenium substituted ferredoxins (see Sect. 8).

We are well aware of the naivety of the whole theoretical approach; it is, however, apparent that smaller J values are needed when the average oxidation state of the iron ions decreases and/or larger delocalization is present. Within this frame, the experimental trends are quite self-consistent. In fact, i) the Mössbauer and NMR spectra within each class of cluster cores are strikingly similar one another, and ii) the gross experimental features are reproduced with sets of J values that make sense, if compared one to the other, in terms of iron-iron distances and expected electron delocalization. These cannot be considered accidental features, and must represent an initial step for the understanding of the electronic structure of all the $[Fe_nS_m]$ cores. At variance with most literature data, we also propose that B is small. Its effect is that of determining the existence of mixed valence pairs. A small B value is enough to cause electronic delocalization, because spin frustration mechanisms are always operative in centers containing more than two iron ions, in which each metal is coupled to all the others.

10 Valence Distribution and Reduction Potentials

One problem only partially addressed up to now is what are the factors affecting the valence distribution. From the $[Fe_2S_2]^+$ case it appears that the electrostatic factors control the electron localization, in the sense that they dictate which iron ion is reduced, just like they are expected to control the overall reduction potential. It has recently been suggested that the partial charges of atoms including H-bonds with the sulfur donor atoms differentiate between Fds and HiPIPs [223]. It has also been suggested that the overall net charges (of ionized residues) account for the observed pattern of reduction potentials of HiPIPs [256]. Other factors like the dipoles induced by the net charges and water-protein interactions are surely very important. Their absolute evaluation at the moment is still very difficult but we expect that understanding of the reduction potential will involve the understanding of the micropotentials of each iron ion, and therefore of the whole valence distribution. By substituting a Cys with a Ser residue, i.e. a sulfur atom with an oxygen atom as donor to one of the iron ions of the $[Fe_4S_4]$ center in HiPIP from *C. vinosum* there is indeed a change in valence distribution, i.e. in the order of micropotentials, and in the whole reduction potential [257].

Similar results have been observed in subsite differentiated models compounds containing the $[Fe_4S_4]^{2+}$ core [66, 258–261]. In these clusters, small variation of redox potential upon coordination change at one of the four iron sites are detected [261]. Furthermore, skewing of electron distribution at the unique subsite toward more ferric or more ferrous character is caused by coordination of either sulfur or nitrogen donor ligands, respectively, indicating a flexible electron density distribution within the cluster core [260].

11 Structural Variations Upon Redox Change

The structural variation upon redox change can be addressed for the case of HiPIP from *C. vinosum* [262, 263] and HiPIP iso-I from *E. halophila* [149, 264], for which complete structure determination of both oxidized and reduced state in solution has been obtained using NMR spectroscopy. Despite the paramagnetism of the proteins, the level of resolution is absolutely comparable to that of diamagnetic proteins of the same size. The structures of the oxidized and reduced proteins are amazingly very similar, as shown in Fig. 16, where the superimposed backbones are shown. It appears that only in some regions there are small variations outside the root mean square deviation [149, 262–264]. Aromatic residues are essentially unaltered. It should be noted that the X-ray structures are more similar to the solution structures of the reduced species. When refinement of solution structures is performed through molecular dynamics approaches, some of the differences between reduced and oxidized protein structures may in part

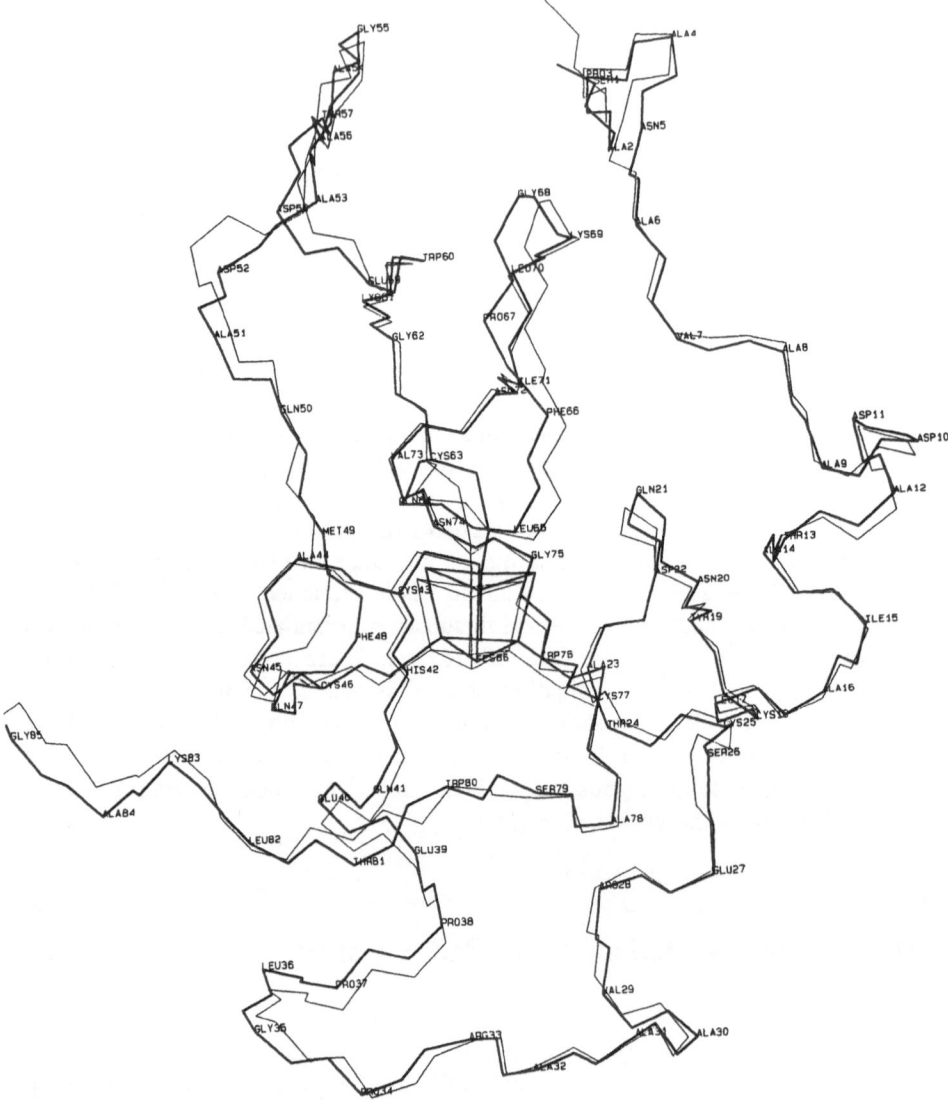

Fig. 16. Superimposed backbones of the structure of reduced (*thin line*) and oxidized (*thick line*, labelled) *C. vinosum* HiPIP in solution, as determined by NMR spectroscopy [262, 263]

depend on the choice of the atomic charges on the cluster atoms. A recent set of values [265] seems to be more appropriate than those previously used [178]. Molecular dynamics investigations have shown relatively high rigidity of this type of protein [178], consistent with the general idea that small reorganizational energies favor fast electron transfer [5]. It is possible therefore that enthalpy variations associated with the electron transfer are small, that the reorganizational energy is small, and that the electrostatic calculations, which provide variations of internal energy, pretty well parallel the $\Delta G^{o\prime}$, and therefore $\Delta E^{o\prime}$ [256]. In the protein structures obtained in solution, indeed, the clusters do not superimpose completely. However, this can be the result of lack of NMR constraints on the FeS cluster, because NMR provides information only on the protein portion.

12 Concluding Remarks

Nowadays, a deep understanding has been reached on the electronic structure of FeS systems. Many properties, in particular the hyperfine coupling of the various nuclei, are understood in terms of a Heisenberg approach whose parameters are largely transferable if corrections for the appropriate oxidation number are taken into account. It is interesting to note that the information obtained through spectroscopy based on hyperfine coupling parameters is related to spins, which we have used as markers for oxidation numbers, whereas the actual picture is such that charges are not integer but largely delocalized.

The next step in the understanding of the electronic structure of iron-sulfur clusters is presumably represented by the development of more accurate theoretical treatments. Extensions of the Heisenberg approach are being pursued, with new Hamiltonians, containing a larger number of parameters and still providing analytical solutions [266–268]. The *ab initio* approach is terribly expensive for open shell ions [222], and not particularly appropriate to obtain spin Hamiltonian parameters. Xα type of calculations have been in progress [164, 269–271], as well as density functional calculations [272], from which an estimate of the value of J is possible. Single crystal ENDOR experiments could allow a more precise determination of the hyperfine constants of the individual iron ions, as already performed on model complexes [87, 233–235].

The procedure to recognize which iron ion in a given oxidation state, as defined above, is bound to a certain cysteine has been established. The future challenge is the understanding of the factors determining valence distribution. This challenge is related to that of understanding the redox properties of the whole polymetallic center. It appears that charges, dipoles, induced dipoles, and solvent accessibility, in other words the whole protein structure, account for every property, from J to B to microscopic and overall potentials. The protein structure, by providing rigidity to the system, also assists the electron transfer process.

We may conclude this article by quoting a comment by Helmut Beinert to one of us: "When I am told that we are able to tell how much of a fraction of an electron goes on which iron ion in the polymetallic center I think that the field of iron-sulfur proteins has gone a long way since the time we used to smell an acidic protein sample to recognize an Fe–S center!"

13 Glossary

AntiCurie Behavior
In NMR, a negative slope of δ *vs.* 1/T (opposite to the **Curie behavior**).

Antiferromagnetic Coupling
The case of **Heisenberg exchange coupling** where the low spin state is the ground state.

Contact Shift
The contribution to the **hyperfine shift** of a nucleus arising from contact coupling (see also **hyperfine coupling**).

Curie Behavior
The 1/T dependence of $\langle S_z \rangle$ and of χ according to Curie law. In NMR of paramagnetic systems the temperature dependence of **hyperfine shifts** follows the Curie law only in the idealized case of an isolated S multiplet. Curie law is not followed quantitatively, for instance, in the presence of zero field splitting of the ground state multiplet. Qualitatively, a positive slope of δ *vs.* 1/T is still termed Curie behavior.

Double Exchange Coupling
In **mixed valence** systems, the exchange coupling mechanism inducing electron delocalization over two centers. In VB language, the double exchange energy is the energy stabilization achieved upon resonance between two limit situations having the extra electron on either of the two centers involved. A consequence of strong double exchange is that the ground state of the system is high spin, as in the presence of Heisenberg **ferromagnetic coupling**.

Ferromagnetic Coupling
The case of **Heisenberg exchange coupling** where the high spin state is the ground state.

Heisenberg Exchange Coupling
The magnetic coupling between unpaired electrons residing on different atoms. As the electron-nucleus **hyperfine coupling**, it is a tensorial interaction; however, is often dominated by a scalar term. In MO language, exchange coupling between two, for instance, $S = \frac{1}{2}$ systems can be described by a bonding and an antibonding MO deriving from the overlap of two half-filled orbitals from the two centers. Depending on the relative magnitude of the bonding-antibonding separation and the electron pairing energy, either the singlet $S' = 0$ or the triplet $S' = 1$ state can be the ground state. The separation between the singlet and the triplet states is the exchange coupling energy (see also **Magnetic Coupling Constant**).

Hyperfine Coupling
The magnetic coupling between electron and nuclear spin magnetic moments. It is customarily divided into dipolar and contact terms, a tensorial quantity the former

and a scalar the latter. In solution, the traceless part of the dipolar tensor vanishes due to rotational averaging, leaving a so-called pseudocontact term besides the contact term. The contact term depends on the presence of finite unpaired electron spin density at the nucleus (see **contact shift**) and the pseudocontact term depends on the **magnetic susceptibility** anisotropy.

Hyperfine Shift
The contribution to the chemical shift δ of a nucleus arising from **hyperfine coupling** with unpaired electrons. It is customarily divided into **contact** and **dipolar shift**. In solution the latter reduces to the trace of the dipolar shift tensor, called **pseudocontact shift**.

Isomer Shift
Energy difference in the absorption of γ-rays by a given nucleus with respect to a reference. It is measured in mm s^{-1} by **Mössbauer spectroscopy**, and it can be used to establish the oxidation state of the element corresponding to the nucleus.

Magnetic Coupling Constant
For a system constituted by two particles whose (electronic or nuclear) spins interact magnetically (with no orbital angular momentum contributions), it is possible to write a spin Hamiltonian of the type in Eq. (1), where J is the exchange integral between centers 1 and 2. J is negative for a ferromagnetic interaction and positive for an antiferromagnetic interaction. J is called the magnetic coupling constant.

Magnetic Susceptibility
The proportionality constant χ between the applied magnetic field and the induced magnetization in a substance. If the magnetization is anisotropic the magnetic susceptibility becomes a tensorial quantity.

Mixed Valence Pair
In a polymetallic system, the metal pair showing electron delocalization over the two centers (see also **double exchange coupling**).

Mössbauer Quadrupole Splitting
A non-spherical electronic distribution in the 1s orbital causes a removal of degeneracy in the $(2I+1)$ nuclear spin levels. For a nucleus with $I = \frac{1}{2}$, such as ^{57}Fe, two levels are generated, and the splitting of the absorption energy is called Mössbauer quadrupole splitting.

Mössbauer Spectroscopy
Measurement of the energy of the nuclear transition resulting from the recoiless absorption of γ-rays. The transition involves changes in the nuclear spin quantum number I. The value of the energy of the nuclear transition depends on the electron density around the nucleus, and it is measured as a shift (**isomer shift**) from a reference sample.

Mulliken Charge
The Mulliken charge is given as the difference between the nuclear charge and the total electron density of the atom. The latter is calculated as the sum of

the squares of the coefficients with which each atomic orbital appears in the molecular orbitals constructed according to the LCAO-MO theory.

Occupation Operator

An electronically delocalized polymetallic system can be described by an Hamiltonian which takes into consideration the fact that magnetic coupling between two metal centers can contribute differently to the total energy depending on which ion the delocalized spin density is considered to reside. To distinguish these situations, an occupation operator is used, which keeps track of the position of the delocalized spin density.

Pseudocontact Shift

The contribution to the **hyperfine shift** of a nucleus arising from **magnetic susceptibility** anisotropy (see also **hyperfine coupling**).

Spin Frustration

In a topologically closed system constituted by (at least) three metal centers with unpaired spin density, in which the main magnetic coupling mechanism is antiferromagnetic, spin frustration occurs: not all the $n(n-1)/2$ possible spin pairs can couple antiferromagnetically, so that **ferromagnetism** is induced.

Subspin

In a polymetallic system constituted by n spins experiencing **Heisenberg exchange coupling**, a total of $n(n-1)/2$ exchange coupling constants J_{ij} can be defined between the individual spins. Analytical solutions for the exchange coupling energy can be in some cases obtained by grouping together all individual spins that experience the same exchange coupling with all other spins. Each group of spins is termed a subspin of the total spin S'.

Superexchange Coupling

A mechanism for exchange coupling between two paramagnetic centers involving filled orbitals of ligands bridging the two centers. It is often the main exchange coupling mechanism, in polymetallic systems.

Unpaired Spin Density

The value of the function

$$\rho = \sum_i [\Psi_i^{\alpha^2}(x, y, z) - \Psi_i^{\beta^2}(x, y, z)]$$

at any point in space of coordinates x, y, z. Ψ_i^{α} and Ψ_i^{β} are the spin-orbital functions for each molecular orbital of the molecule, and the difference of their squares gives the unbalance between the two spin densities at that point. Integration of the expression in parenthesis over space gives zero for filled orbitals and unity for half-filled orbitals. Integration of ρ over space gives the number of unpaired electrons of the molecule, 2S. Unpaired spin density at a nucleus is given by

$$\rho = \sum_i [\Psi_i^{\alpha^2}(0) - \Psi_i^{\beta^2}(0)]$$

where the summation needs only to be performed over the s orbitals of the atom which the nucleus belongs to.

14 References

1. Mortenson LE, Valentine RC, Carnahan JC (1962) Biochem Biophys Res Commun 7: 448
2. Arnon DI (1965) Science 149: 1460
3. Hochkoeppler A, Ciurli S, Venturoli G, Zannoni D (1995) FEBS Lett 357: 70
4. Frausto da Silva JJR, Williams RJP (1991) The biological chemistry of the elements. Oxford University Press, Oxford
5. Gray HB, Elllis WR (1994) In: Bertini I, Gray HB, Lippard SJ, Valentine JS (eds) Bioinorganic chemistry. University Science Books, Mill Valley, CA, p 315
6. (a) Beinert H, Kennedy M (1989) Eur J Biochem 186: 5
   (b) Emptage H (1988) In: Que L, Jr (ed) Metal clusters in proteins. ACS Symposium Series 372, American Chemical Society, Washington, DC, Chpt 17
7. Kutchta RD, Hanson GR, Holmquist B, Abeles RH (1986) Biochemistry 25: 7301
8. Cunningham RP, Asahara H, Bank JF, Scholes CP, Salerno JC, Surerus KK, Münck E, McCracken J, Peisach J, Emptage MH (1989) Biochemistry 28: 4450
9. Grandoni JA, Switzer RL, Makaroff CA, Zalkin H (1989) J Biol Chem 264: 6058
10. Vollmer SJ, Debrunner PG (1983) J Biol Chem 258: 14284
11. Flint DH, Emptage MH (1988) J Biol Chem 263: 3558
12. Dunham WR, Palmer G, Sands RH, Bearden AJ (1971) Biochim Biophys Acta 253: 373
13. Fukuyama K, Hase T, Matsumoto S, Tsukihara T, Katsube Y, Tanaka N, Kakudo M, Wada K, Matsubara H (1980) Nature (London) 286: 522
14. Tsukihara T, Fukuyama K, Nakamura M, Katsube Y, Tanaka N, Kakudo M, Wada K, Hase T, Matsubara H (1981) J Biochem 90: 1763
15. Tsutsui T, Tsukihara T, Fukuyama K, Katsube Y, Hase T, Matsubara H, Nishikawa Y, Tanaka N (1983) J Biochem 94: 299
16. Rypniewski WR, Breiter DR, Benning MM, Wesenberg G, Oh B-H, Markley JL, Rayment I, Holden HM (1991) Biochemisty 30: 4126
17. Dunham WR, Bearden AJ, Salmeen I, Palmer G, Sands RH, Orme-Johnson WH, Beinert H (1971) Biochim Biophys Acta 253: 134
18. Carter CW, Jr, Freer ST, Xuong NH, Alden RA, Kraut J (1971) Cold Spring Harbor Symp Quant Biol 36: 381
19. Sieker LC, Adman ET, Jensen LH (1972) Nature (London) 235: 40
20. Carter CW, Jr, Kraut J, Freer ST, Alden RA, Sieker LC, Adman E, Jensen LH (1972) Proc Natl Acad Sci USA 69: 3526
21. Adman ET, Sieker LC, Jensen LH (1973) J Biol Chem 248: 3987
22. Carter CW, Jr, Kraut J, Freer ST, Alden RA (1974) J Biol Chem 249: 6339
23. Carter CW, Jr, Kraut J, Freer ST, Xuong NH, Alden RA, Bartsch RG (1974) J Biol Chem 249: 4212
24. Adman ET, Sieker LC, Jensen LH (1976) J Biol Chem 251: 3801
25. Freer ST, Alden RA, Carter CW, Jr, Kraut J (1975) J Biol Chem (1975) 250: 46
26. Stout GH, Turley S, Sieker LC, Jensen LH (1988) Proc Natl Acad Sci USA 85: 1022
27. Fukuyama K, Nagahara Y, Tsukihara T, Katsube Y (1988) J Mol Biol 199: 183
28. Fukuyama K, Matsubara H, Tsukihara T, Katsube Y (1989) J Mol Biol 210: 383
29. Stout CD (1989) J Mol Biol 205: 545
30. Breiter DR, Meyer TE, Rayment I, Holden HM (1991) J Biol Chem 266: 18660
31. Rayment I, Wesenberg G, Meyer TE, Cusanovich MA, Holden HM (1992) J Mol Biol 228: 672
32. Benning MM, Meyer TE, Rayment I, Holden HM (1994) Biochemistry 33: 2476
33. Dueé ED, Fanchon E, Vicat J, Sieker LC, Meyer J, Moulis J-M (1994) J Mol Biol 243: 683
34. Sery A, Housset D, Serre L, Bonicel J, Hutchikian C, Frey M, Roth AM (1994) Biochemistry 33: 15408

35. Kim J, Rees DC, (1992) Nature 360: 553
36. Bolin JT et al. (1993) in: Stiefel E et al. (eds) Molybdenum enzymes, cofactors and models. American Chemical Society, Washington, DC. 186–195
37. DePamphilis BV, Averill BA, Herskovitz T, Que L, Jr, Holm RH (1974) J Am Chem Soc 96: 4159
38. Kissinger CR, Adman ET, Sieker LC, Jensen LH (1988) J Am Chem Soc 110: 8721
39. Kissinger CR, Adman ET, Sieker LC, Jensen LH, LeGall J (1989) FEBS Lett 244: 447
40. Robbins AH, Stout CD (1989) Proc Natl Acad Sci USA 86: 3639
41. Huynh BH, Moura JJG, Moura I, Kent TA, LeGall J, Xavier AV, Münck E (1980) J Biol Chem 255: 3242
42. Dunham WR, Carroll RT, Thompson JF, Sands RH, Funk MO (1990) Eur J Biochem 190: 611
43. Pierik AJ, Wolbert RBG, Mutsaers PHA, Hagen WR, Veeger C (1992) Eur J Biochem 206: 697
44. Pierik AJ, Hagen WR, Dunham WR, Sands RH (1992) Eur J Biochem 206: 705
45. Moura I, Taveres P, Moura JJG, Ravi N, Huynh BH, Liu M-Y, LeGall J (1992) J Biol Chem 267: 4489
46. Kanatzidis MG, Hagen WR, Dunham WR, Lester RK, Coucouvanis D (1985) J Am Chem Soc 107: 953
47. Kanatzidis MG, Salifoglou A, Coucouvanis D (1985) J Am Chem Soc (1985) 107: 3358
48. Coucouvanis D, Kanatzidis MG, Dunham WR, Hagen WR (1984) J Am Chem Soc 106: 7998
49. Saak W, Henkel G, Pohl S (1984) Angew Chem Int Ed Engl 23: 150
50. Kanatzidis MG, Dunham WR, Hagen WR, Coucouvanis D (1984) J Chem Soc Chem Commun 356–358
51. Saak W, Pohl S (1985) Z. Naturforsch 40b: 1105
52. Kanatzidis MG, Salifoglou A, Coucouvanis D (1986) Inorg Chem 25: 2460
53. Herriot JR, Sieker LC, Jensen LH, Lovenberg W (1970) J Mol Biol 50: 391
54. Watenpaugh KD, Sieker LC, Herriot JR, Jensen LH (1971) Cold Spring Harbor Symp Quant Biol 36: 359
55. Watenpaugh KD, Sieker LC, Herriot JR, Jensen LH (1973) Acta Crystallogr Sect B, 29: 943
56. Adman ET, Sieker LC, Jensen LH, Bruschi M, LeGall J (1977) J Mol Biol 112: 113
57. Watenpaugh KD, Sieker LC, Jensen LH (1979) J Mol Biol 131: 509
58. Sieker LC, Stenkamp RE, Jensen LH, Prickril B, LeGall J (1986) FEBS Lett. 208: 73
59. Frey M, Sieker LC, Payan F, Haser R, Bruschi M, Pepe G, LeGall J (1987) J Mol Biol 197: 525
60. Ibers JA, Holm RH (1980) Science 209: 223
61. Berg JM, Holm RH (1982) In: Spiro TG (ed) Iron-sulfur proteins. Wiley-Interscience, New York
62. O'Sullivan T, Millar M (1985) J Am Chem Soc 107: 4096
63. Mascharak PK, Papaefthymiou GC, Frankel RB, Holm RH (1981) J Am Chem Soc 103: 6110
64. Thomson AJ (1985) In: Harrison P (ed) Mettalloproteins. VCH, Weinheim p 79
65. Beinert H (1990) FASEB Journal 4: 2483
66. Holm RH, Ciurli S, Weigel JA (1990) Progr Inorg Chem 38: 1
67. Howard JB, Rees DC (1991) Adv Protein Chem 42: 199
68. Holm RH (1992) In: Cammack R (ed) Adv Inorg Chem Academic Press, San Diego (CA) Vol 38, p 1
69. Cammack R (1992) In: Cammack R (ed) Adv Inorg Chem Academic Press, San Diego (CA) Vol 38, p 281
70. Stevens KWH (1985) In: Willett RD, Gatteschi D, Kahn O (eds) Magneto-structural correlations in exchange-coupled systems. Reidel, Dordrecht, p 105
71. Hay PJ, Thibeault JC, Hoffmann R (1975) J Am Chem Soc 97: 4884
72. Mukherjee RN, Stack TDP, Holm RH (1988) J Am Chem Soc 110: 1850
73. Bencini A, Gatteschi D (1990) In: Electron paramagnetic resonance of exchange-coupled systems, VCH, Berlin
74. Palmer G, Dunham WR, Fee JA, Sands RH, Itzuka T, Yonetani T (1971) Biochim Biophys Acta 245: 201
75. Anderson RE, Dunham WR, Sands RH, Bearden AJ, Crespi HL (1975) Biochim Biophys Acta 408: 306
76. Gillum WO, Frankel RB, Foner S, Holm RH (1976) Inorg Chem 15: 1095
77. Sinn E (1970) Coord Chem Reviews 5: 313

78. Salmeen IT, Palmer G (1972) Arch Biochem Biophys 150: 767
79. Banci L, Bertini I, Luchinat C (1970) Struct Bonding 71: 113
80. Skjeldal L, Markley JL, Coghlan VM, Vickery LE (1991) Biochemistry 30: 9078
81. Banci L, Bertini I, Luchinat C (1991) Nuclear and electron relaxation. VCH, Weinheim
82. Luchinat C, Ciurli S (1993) In: Berliner LJ, Reuben J (eds) Biological magnetic resonance Plenum Press, New York 1993, 357–420
83. Slichter CP (1955) Phys Rev 99: 479
84. Benini S, Ciurli S, Luchinat C (1995) Inorg Chem 34: 417
85. Werth MT, Kurtz DM, Moura I, LeGall J (1987) J Am Chem Soc 109: 273
86. Hagen KS, Watson AD, Holm RH (1983) J Am Chem Soc 105: 3905
87. (a) Mouesca JM, Rius G, Lamotte B (1993) J Am Chem Soc 115: 4714 (b) Gloux J, Gloux P, Lamotte B, Mouesca JM, Rius G (1994) J Am Chem Soc 116: 1953
88. Phillips WD, Poe M, Weiher JF, McDonald CC, Lovenberg W (1970) Nature (London) 227: 574
89. Schultz C, Debrunner PG (1976) J Phys (Paris) C6 (37): 153
90. Rao KK, Evans MCW, Cammack R, Hall DO, Thompson CL, Jackson PJ, Johnson CE (1972) Biochem J 129: 1063
91. Moura I, Huynh BH, Hausinger RP, LeGall J, Xavier AV, Münck E (1980) J Biol Chem 255: 2493
92. Kostikas A, Petrouleas V, Simopoulos A, Coucouvanis D, Holah DG (1976) Chem Phys Lett 38: 582
93. Petrouleas V, Simopoulos A, Kostikas A, Coucouvanis D (1976) J Phys (Paris) C6 (37): 159
94. Lane RW, Ibers JA, Frankel RB, Papaefthymiou GC, Holm RH (1977) J Am Chem Soc 99: 84
95. Frankel RB, Papaefthymiou GC, Lane RW, Holm RH (1976) J Phys (Paris) C6 (37): 165
96. Rao KK, Cammack R, Hall DO, Johnson CE (1971) Biochem J 122: 257
97. Johnson CE, Elstner E, Gibson JF, Benfield G, Evans MCW, Hall DO (1968) Nature 220: 1291
98. Münck E, Debrunner PG, Tsibris JCM, Gunsalus IC (1972) Biochemistry 11: 855
99. Cammack R, Rao KK, Hall DO, Johnson CE (1971) Biochem J 125: 849
100. Geary PJ, Dickson DPE (1981) Biochem J 195: 199
101. Fee JA, Findling KL, Yoshida T, Hille R, Tarr GE, Hearshen DO, Dunham WR, Day EP, Kent TA, Münck E (1984) J Biol Chem 259: 124
102. Bill E, Bernhardt F-H, Trautwein AX (1981) Eur J Biochem 121: 39
103. Papaefthymiou V, Girerd J-J, Moura I, Moura JJG, Münck E (1987) J Am Chem Soc 109: 4703
104. Emptage MH, Kent TA, Huynh BH, Rawlings J, Orme-Johnson WH, Münck E (1980) J Biol Chem 255: 1793
105. Kent TA, Dreyer J-L, Kennedy MC, Huynh BH, Emptage MH, Beinert H, Münck E (1982) Proc Natl Acad Sci USA 79: 1096
106. Kent TA, Emptage MH, Merkle H, Kennedy MC, Beinert H, Münck E (1985) J Biol Chem 260: 6871
107. Surerus KK, Kennedy MC, Beinert H, Münck E (1989) Proc Natl Acad Sci USA 86: 9846
108. Hille R, Yoshida T, Tarr GE, Williams CH, Ludwig MH, Fee JA, Kent TA, Huynh BH, Münck E (1983) J Biol Chem 258: 13008
109. Srivastava KKP, Surerus KK, Conover RC, Johnson MK, Park J-B, Adams MWW, Münck E (1993) Inorg Chem 32: 927
110. Weigel JA, Holm RH, Surerus KK, Münck E (1989) J Am Chem Soc 111: 9246
111. Weigel JA, Srivastava KKP, Day EP, Münck E, Holm RH (1990) J Am Chem Soc 112: 8015
112. Cammack R (1976) J Phys (Paris) C6 (37): 137
113. Middleton P, Dickson DPE, Johnson CE, Rush JD (1978) Eur J Biochem 88: 135
114. Mullinger RN, Cammack R, Rao KK, Hall DO, Dickson DPE, Johnson CE, Rush JD, Simopoulos A (1975) Biochem J 151: 75
115. Dickson DPE, Johnson CE, Middleton P, Rush JD, Cammack R, Hall DO, Mullinger RN, Rao KK (1976) J Phys (Paris) C6(37): 171
116. Thompson CL, Johnson CE, Dickson DPE, Cammack R, Hall DO, Weser U, Rao KK (1974) Biochem J 139: 97
117. Gersonde K, Schlaak H-E, Breitenbach M, Parak F, Eicher H, Zgorzalla W, Kalvius MG, Mayer A (1974) Eur J Biochem 43: 307
118. Moss TH, Bearden AJ, Bartsch RG, Cusanovich MA, San Pietro A, (1968) Biochemistry 7: 1591

119. Moura JJG, Moura I, Kent TA, Lipscomb JD, Huynh BH, LeGall J, Xavier AV, Münck E (1982) J Biol Chem 257: 6259
120. Emptage MH, Kent TA, Kennedy MC, Beinert H, Münck E (1983) Proc Natl Acad Sci USA 80: 4674
121. Middleton P, Dickson DPE, Johnson CE, Rush JD (1980) Eur J Biochem 104: 289
122. Dickson DPE, Johnson CE, Cammack R, Evans MCW, Hall DO, Rao KK (1974) Biochem J 139: 105
123. Dickson DPE, Cammack R (1974) Biochem J 143: 763
124. Bertini I, Campos AP, Luchinat C, Teixeira M (1993) J Inorg Biochem 52: 227
125. Papaefthymiou V, Millar MM, Münck E (1986) Inorg Chem 25: 3010
126. Lindahl PA, Day EP, Kent TA, Orme-Johnson WH, Münck E (1985) J Biol Chem 260: 11160
127. Frankel RB, Averill BA, Holm RH (1976) J Phys (Paris) C6(37): 107
128. Holm RH, Averill BA, Frankel RB, Gray HB, Siiman O, Grunthaner FJ (1974) J Am Chem Soc 96: 2644
129. Laskowski EJ, Frankel RB, Gillum WO, Papaefthymiou GC, Renaud J, Ibers JA, Holm RH (1978) J Am Chem Soc 100: 5322
130. Mayerle JJ, Frankel RB, Holm RH, Ibers JA, Phillips WD, Weiher JF (1973) Proc Natl Acad Sci USA 70: 2429
131. Mayerle JJ, Denmark SE, DePamphilis BV, Ibers JA, Holm RH (1975) J Am Chem Soc 97: 1032
132. Reynolds JG, Holm RH (1980) Inorg Chem 19: 3257
133. (a) Yachandra VK, Hare J, Gewirth A, Czernuszewicz RS, Kimura T, Holm RH, Spiro TG (1983) J Am Chem Soc 105: 6462 (b) Backes G, Mino Y, Loehr TM, Meyer TE, Cusanovich MA, Sweeney WV, Adman ET, Loehr JS (1991) J Am Chem Soc 113: 2055
134. Wolff TE, Berg JM, Hodgson KO, Frankel RB, Holm RH (1979) J Am Chem Soc 101: 4140
135. Christou G, Mascharak PK, Armstrong WH, Papaefthymiou GC, Frankel RB, Holm RH (1982) J Am Chem Soc 104: 2820
136. Tsibris JCM, Woody RW (1970) Coord Chem Rev 5: 417
137. Holm RH, Phillips WD, Averill BA, Mayerle JJ, Herskowitz T (1974) J Am Chem Soc 96: 2109
138. Nettesheim DG, Meyer TE, Feinberg BA, Otvos JD (1983) J Biol Chem 258: 8235
139. Krishnamoorthy R, Markley JL, Cusanovich MA, Przysiecki CT, Meyer TE (1986) Biochemistry 25: 60
140. Krishnamoorthy R, Cusanovich MA, Meyer TE, Przysiecki CT (1989) Eur J Biochem 181: 81
141. Sola M, Cowan JA, Gray HB (1989) Biochemistry 28: 5261
142. Bertini I, Briganti F, Luchinat C, Scozzafava A, Sola M (1991) J Am Chem Soc 113: 1237
143. Banci L, Bertini I, Briganti F, Luchinat C, Scozzafava A, Vicens-Oliver M (1991) Inorg Chem 30: 4517
144. Bertini I, Capozzi F, Luchinat C, Piccioli M, Vicens-Oliver M (1992) Inorg Chim Acta 198–200: 483
145. Gaillard J, Albrand J-P, Moulis J-M, Wemmer DE (1992) Biochemistry 31: 5632
146. Bertini I, Capozzi F, Ciurli S, Luchinat C, Messori L, Piccioli M (1992) J Am Chem Soc 114: 3332
147. Bertini I, Capozzi F, Luchinat C, Piccioli M (1993) Eur J Biochem 212: 69
148. Bertini I, Felli I, Kastrau DHW, Luchinat C, Piccioli M, Viezzoli MS (1994) Eur J Biochem 225: 703
149. Banci L, Bertini I, Eltis LD, Felli I, Kastrau DHW, Luchinat C, Piccioli M, Pierattelli R, Smith M (1994) Eur J Biochem 225: 715
150. Bertini I, Gaudemer A, Luchinat C, Piccioli M (1993) Biochemistry 32: 12887
151. Ciurli S, Cremonini MA, Kofod P, Luchinat C submitted
152. Poe M, Phillips WD, McDonald CC, Lovenberg W (1970) Proc Natl Acad Sci USA 65: 797
153. Packer EL, Sweeney WV, Rabinowitz JC, Sternlich H, Shaw EN (1977) J Biol Chem 252: 2245
154. Bertini I, Briganti F, Luchinat C, Scozzafava A (1990) Inorg Chem 29: 1874
155. Bertini I, Briganti F, Luchinat C, Messori L, Monnanni R, Scozzafava A, Vallini G (1991) FEBS Lett 289: 253
156. Busse SC, LaMar GN, Howard JB (1991) J Biol Chem 266: 23714
157. Bertini I, Briganti F, Luchinat C, Messori L, Monnanni R, Scozzafava A, Vallini G (1992) Eur J Biochem 204: 831
158. Bertini I, Capozzi F, Luchinat C, Piccioli M, Vila AJ (1994) J Am Chem Soc 116: 651

159. Donaire A, Gorst CM, Zhou ZH, Adams MWW, LaMar GN (1994) J Am Chem Soc 116: 6841
160. Zener C (1951) Phys Rev 82: 403
161. Anderson PW, Hasegawa H (1955) Phys Rev 100: 675
162. Karpenko BV (1976) J Magnetism Magnetic Materials 3: 267
163. Belinskii MI, Tsukerblat BS, Gerbeleu NV (1983) Sov Phys Solid State 25: 497
164. Noodleman L, Baerends EJ (1984) J Am Chem Soc 106: 2316
165. Noodleman L (1988) Inorg Chem 27: 3677
166. Jordanov J, Roth EKH, Fries PH, Noodleman L (1990) Inorg Chem 29: 4288
167. Noodleman L (1991) Inorg Chem 30: 256
168. Noodleman L (1991) Inorg Chem 30: 246
169. Borshch SA, Chibotaru LF (1989) Chem Phys 135: 375
170. Blondin G, Girerd J-J (1990) Chem Rev 90: 1359
171. Drüecke S, Chaudhuri P, Pohl K, Wieghardt K, Ding X-Q, Bill E, Sawaryn A, Trauwein AX, Winkler H, Gurman SJ (1989) J Chem Soc Chem Commun 59
172. (a) Ding X-Q, Bominaar EL, Bill E, Winkler H, Trautwein AX, Drüecke S, Chauduri P, Wieghardt KJ (1990) Chem Phys 92: 178. (b) Trautwein AX, Bill E, Ding X-Q, Winkler H, Kostikas A, Papaefthymiou V, Simopoulos A, Beardwood P, Gibson JF (1993) J Inorg Biochem 51: 481
173. Nettesheim DG, Harder SR, Feinberg BA, Otvos JD (1992) Biochemistry 31: 1234
174. Banci L, Bertini I, Capozzi F, Carloni P, Ciurli S, Luchinat C, Piccioli M (1993) J Am Chem Soc 115: 3431
175. Banci L, Bertini I, Ciurli S, Ferretti S, Luchinat C, Piccioli M (1993) Biochemistry 32: 9387
176. Bertini I, Capozzi F, Eltis L D, Felli I, Luchinat C, Piccioli M, Inorg Chem in press
177. Cowan JA, Sola M (1990) Biochemistry 29: 5633
178. Banci L, Bertini I, Carloni P, Luchinat C, Orioli PL (1992) J Am Chem Soc 114: 10683
179. Karplus M (1959) J Chem Phys 30: 11
180. Karplus M (1963) J Am Chem Soc 85: 2870
181. Fitzgerald RJ, Drago RS (1968) J Am Chem Soc 90: 2523
182. Ho FF-L, Reilly CN (1969) Anal Chem 41: 1835
183. Pratt L, Smith BB (1969) Trans Faraday Soc 65: 915
184. Karplus M, Fraenkel GK (1961) J Chem Phys 35: 1312
185. Heller C, McConnell HM (1960) J Chem Phys 32: 1535
186. Stone EW, Maki AH (1962) J Chem Phys 37: 1326
187. Huber G, Gaillard J, Moulis J-M in press
188. Shetna YI, Dervartalian DV, Beinert H (1968) Biochem Biophys Res Commun 31: 862
189. Salerno JC, Ohnishi T, Blum H, Leigh JS (1977) Biochim Biophys Acta 494: 191
190. Gayda J-P, Gibson JF, Cammack R, Hall DO, Mullinger R (1976) Biochim Biophys Acta 434: 154
191. Gayda J-P, Bertrand P, More C, Cammack R (1981) Biochimie 63: 847
192. Bertrand P, Guigliarelli B, More C (1991) New J Chem 15: 445
193. Blondin G, Borshch S, Girerd J-J (1992) Comm Inorg Chem 12: 315
194. Belinskii M, Bertini I, Galas O, Luchinat C (1995) Z Naturforsch 50a: 75
195. Peisach J, Blumberg WE, Lode ET, Coon MJ (1971) J Biol Chem 246: 5877
196. Rao KK, Smith RV, Cammack R, Evans MCW, Hall DO, Johnson CE (1972) Biochem J 129: 1159
197. Hall DO, Evans MCW (1969) Nature (London) 223: 1342
198. Fritz J, Anderson R, Fee J, Palmer G, Sands RH, Tsibris JCM, Gunsalus IC, Orme-Johnson WH, Beinert H (1971) Biochim Biophys Acta 253: 110
199. Siedow JN, Power S, de la Rosa FF, Palmer G (1978) J Biol Chem 263: 2392
200. Prince RC (1983) Biochim Biophys Acta 723: 133
201. Axcell BC, Geary PJ (1975) Biochem J 146: 173
202. Twilfer H, Bernhardt F-H, Gersonde K (1981) Eur J Biochem 119: 595
203. Bernhardt F-H, Heymann E, Traylor PS (1978) Eur J Biochem 92: 209
204. Sauber K, Fröhner C, Rosenberg G, Eberspächer J, Lingens F (1977) Eur J Biochem 74: 89
205. Moura I, Moura JJG, Huynh BH, Santos H, LeGall J, Xavier AV (1982) Eur J Biochem 126: 95
206. Antanaitis BC, Moss TH (1975) Biochim, Biophys Acta 405: 262
207. Dunham WR, Hagen WR, Fee JA, Sands RH, Dunbar JB, Humblet C (1991) Biochim Biophys Acta 1079: 253
208. Beinert H, Thomson AJ (1983) Arch Biochem Biophys 222: 333

209. Moulis J-M, Scherrer N, Gagnon J, Forest E, Petillot Y, Garcia D (1993) Arch Biochem Biophys 305: 186
210. Anderson RE, Anger G, Petersson L, Ehrenberg A, Cammack R, Hall DO, Mullinger R, Rao KK (1975) Biochim Biophys Acta 376: 63
211. Stombaugh NA, Burris RH, Orme-Johnson WH (1973) J Biol Chem 22: 7951
212. Mathews R, Charlton S, Sands RH, Palmer G (1974) J Biol Chem 249: 4326
213. Yoch DC, Carithers RP, Arnon DI (1977) J Biol Chem 252: 7453
214. Gibson JF, Hall DO, Thornley JHM, Whatley FR (1966) Proc Natl Acad Sci USA 56: 987
215. Thornley JHM, Gibson JF, Whatley FR, Hall DO (1966) Biochem Biophys Res Commun 24: 877
216. Bertrand P, Gayda J-P (1979) Biochim Biophys Acta 579: 107
217. Bertrand P, Guigliarelli B, Gayda J-P, Beardwood P, Gibson JF (1985) Biochim Biophys Acta 831: 261
218. Bertini I, Lanini G, Luchinat C (1984) Inorg Chem 23: 2729
219. Dugad LB, LaMar GN, Banci L, Bertini I (1990) Biochemistry 29: 2263
220. Skjeldal L, Westler WM, Oh B-H, Krezel AM, Holden HM, Jacobson BL, Rayment I, Markley JL (1991) Biochemistry 30: 7363
221. Matsubara H, Hase T, Wakabayashi S, Wada K (1981) In: Sigman DS, Brazier MAB (eds) Evolution of protein structure and function, Acamedic Press, New York,
222. Carloni P, Corongiu G submitted
223. Langen R, Jensen GM, Jacob U, Stephens PJ, Warshel A (1992) J Biol Chem 267: 25625
224. Cheng H, Xia B, Reed GH, Markley JL (1994) Biochemistry 33: 3155
225. Kent TA, Huynh BH, Münck E (1980) Proc Natl Acad Sci USA 77: 6574
226. Day EP, Peterson J, Bonvoisin JJ, Moura I, Moura JJG (1988) J Biol Chem 263: 3684
227. Busse SC, LaMar GN, Yu LP, Howard JB, Smith ET, Zhou ZH, Adams MWW (1992) Biochemistry 31: 11952
228. Macedo AL, Moura I, Moura JJG, LeGall J, Huynh BH (1993) Inorg Chem 32: 1101
229. Gayda J-P, Bertrand P, Theodule F-X, Moura JJG (1982) J Chem Phys 77: 3387
230. Guigliarelli B, Gayda J-P, Bertrand P, More C (1986) Biochim Biophys Acta 871: 149
231. Bertini I, Briganti F, Calzolai L, Messori L, Scozzafava A (1993) FEBS Lett 332: 268
232. Benelli B, Bertini I, Capozzi F, Luchinat C (1995) Gazz Chim Ital 124: 469
233. Gloux J, Gloux P, Lamotte B, Rius G (1985) Phys Rev Lett 54: 599
234. Rius G, Lamotte B (1989) J Am Chem Soc 111: 2464
235. Mouesca JM, Lamotte B, Rius G (1991) J Inorg Biochem 43: 251
236. Banci L, Bertini I, Briganti F, Scozzafava A, Vicens-Oliver M, Luchinat C (1991) Inorg Chim Acta 180: 171
237. Banci L, Bertini I, Briganti F, Luchinat C (1991) New J Chem 15: 467
238. Przysiecki CT, Meyer TE, Cusanovich MA (1985) Biochemistry 24: 2542
239. Bertini I, Ciurli S, Dikiy A, Luchinat C (1993) J Am Chem Soc 115: 12020
240. Johnson MK, Thomson AJ, Robinson AE, Rao KK, Hall DO (1981) Biochim Biophys Acta 667: 433
241. Hales BJ, Langosch J, Case EE (1986) J Biol Chem 261: 15301
242. Lindahl PA, Gorelick NJ, Münck E, Orme-Johnson WH (1987) J Biol Chem 262: 14945
243. Mai X, Adams MWW (1994) J Biol Chem 269: 16726
244. Carney MJ, Holm RH, Papaefthymiou GC, Frankel RB (1986) J Am Chem Soc 108: 3519
245. Carney MJ, Papaefthymiou GC, Spartalian K, Frankel RB, Holm RH (1988) J Am Chem Soc 110: 6084
246. Carney MJ, Papaefthymiou GC, Whitener MA, Spartalian K, Frankel RB, Holm RH (1988) Inorg Chem 27: 346
247. Moulis JM, Auric P, Gaillard J, Meyer J (1984) J Biol Chem 259: 11396
248. Gaillard J, Moulis J-M, Auric P, Meyer J (1986) Biochemistry 25: 464
249. Auric P, Gaillard J, Meyer J, Moulis J-M (1987) Biochem J 242: 525
250. Gaillard J, Moulis J-M, Meyer J (1987) Inorg Chem 26: 320
251. George SJ , Thomson AJ, Crabtree DE, Meyer J, Moulis J-M (1991) New J Chem 15: 455
252. Phillips WD, McDonald CC, Stombaugh NA, Orme-Johnson WH (1974) Proc Natl Acad Sci USA 71: 140
253. Nagayama K, Ozaki Y, Kyogoku Y, Hase T, Matsubara H (1983) J Biochem 94: 893
254. Thomson AJ, Robinson AE, Johnson MK, Moura JJG, Moura I, Xavier AV, LeGall J (1981) Biochim Biophys Acta 670: 93
255. Bertini I, Briganti F, Luchinat C (1990) Inorg Chim Acta 175: 9

256. Banci L, Bertini I, Ciurli S, Luchinat C, Pierattelli R, Inorg Chim Acta in press
257. Babini E, Bertini I, Borsari M, Capozzi F, Dikiy A, Eltis D, Luchinat C submitted
258. Stack TDP, Holm RH (1987) J Am Chem Soc 109: 2546
259. Stack TDP, Holm RH (1988) J Am Chem Soc 110: 2484
260. Ciurli S, Carrié M, Weigel JA, Carney MJ, Stack TDP, Papefthymiou GC, Holm RH (1990) J Am Chem Soc 112: 2654
261. Weigel JA, Holm RH (1991) J Am Chem Soc 113: 4184
262. Banci L, Bertini I, Dikiy A, Kastrau DHW, Luchinat C, Sompornipisut P (1995) Biochemistry 34: 206 in press
263. Bertini I, Dikiy A, Kastrau DHW, Luchinat C, Sompornpisut P submitted
264. Bertini I et al., submitted
265. Mouesca JM, Chen JL, Noodleman L, Bashford D, Case DA (1994) J Am Chem Soc 116: 11898
266. Belinskii M (1993) Chem Phys 172: 189
267. Belinskii M (1993) Chem Phys 172: 213
268. Belinskii M (1993) Chem Phys 173: 27
269. Norman JC, Jr, Bryan PB, Noodleman L (1980) J Am Chem Soc 102: 4279
270. Aizman A, Case DA (1982) J Am Chem Soc 104: 3269
271. Noodleman L, Norman JG, Jr, Osborne JH, Aizman A, Case DA (1985) J Am Chem Soc 107: 3418
272. Mouesca JM, Chen JL, Noodlemann L, Bashford D, Case DA (1994) J Am Chem Soc 116: 11898

On Helices Resulting from a Cooperative Jahn-Teller Effect in Hexagonal Perovskites

Wim J.A. Maaskant

Leiden Institute of Chemistry, Gorlaeus Laboratories, University of Leiden, P.O.-Box 9502, 2300 RA Leiden, The Netherlands

Based on the observations of the cooperative $E \otimes \varepsilon$ Jahn–Teller effect (JTE) in compounds with the hexagonal perovskite structure (ABX_3 2L compounds with an h-stacking) the occurrence and non-occurrence of helical structures as e.g. in $CsCuCl_3$ and $[N(CH_3)_4]CuCl_3$ ($TMCuCl_3$) are discussed. In particular the relative magnitude of the radii of Rb^+ and Cs^+ with respect to the radius of Cl^- can explain the difference in structure of $RbCuCl_3$ and $CsCuCl_3$. This can be shown by comparing new neutron diffraction measurements of the acoustic phonon modes in $CsFeCl_3$ with those previously determined for $RbFeCl_3$. The acoustic phonon branches also demonstrate the phenomenon of structural resonance, which leads to energy lowering in cases of formation of helices. It is further shown that, in order to form a structural helix, displacements of ions related to ferro-electricity as well as to ferro-elasticity are necessary. The quadrupolar distortions resulting from the $E \otimes \varepsilon$ JTE are therefore complemented with dipolar shifts of layers perpendicular to the c-axis of the structure of $CsCuCl_3$. The structure of $CsCuBr_3$ is also shown to belong to this family of helices. It arises through distortion waves at the A-point in the first Brillouin zone and changes from an h-stacking to an hc-stacking of layers. Although a phase transition between the chloride and the bromide has not been observed, the deformation in $CsCuBr_3$ reminds us of a Spin–Peierls transition. A new mechanism connected with ferromagnetic exchange along the columns of octahedra is proposed to explain the structure of β-$RbCrCl_3$ and the low temperature structures (space group C2) of $CsCrCl_3$ and $RbCrCl_3$. The larger radius of Cs^+ can explain why the β-$RbCrCl_3$ structure does not occur for $CsCrCl_3$. The structures having the C2 spacegroup resemble racemic dl-compounds. The question about the stability of the helix in β-$CsCuCl_3$ remains difficult to answer without calculations. As experiment shows, other influences can easily disturb its formation. Probably because of the smaller radius of Cu^{2+} with respect to Cr^{2+}, the electrostatic energy stabilizes β-$CsCuCl_3$ with respect to γ-$CsCrCl_3$.

1 Introduction

In this paper we discuss structures of ABX_3-compounds having the aristotype structure of the so-called hexagonal perovskites (Fig. 1), where B^{2+} exhibits the $E \otimes \varepsilon$ Jahn–Teller effect. Here A is a large monopositive ion like Rb^+, Cs^+ or tetramethylammonium (TM^+). X is Cl^-, Br^-, or J^-. B can be one of the following non-Jahn–Teller ions: Mg^{2+}, V^{2+}, Fe^{2+}, Ni^{2+}, Mn^{2+}, Cd^{2+}, but our interest is in the Cu^{2+} and Cr^{2+}-compounds. Some of the Cu-compounds exhibit a helical distortion with respect to the aristotype. Well-known are the room temperature structure of $CsCuCl_3$ [1], and one of the structures of $TMCuCl_3$ [2, 3]. We present evidence that the room temperature structure of $CsCuBr_3$ [4] also belongs to the helical compounds. Apparently the Jahn–Teller effect (JTE) is the cause of these helical structures. However, no two helical structures are similar. And why doesn't $RbCuCl_3$ form a helix? Similarly for the Cr-compounds, why are these different from $CsCuCl_3$?

These hexagonal perovskites differ in structure from the well-known cubic perovskites. The larger ions A^+ and X^- form (almost) close packed layers, which are hexagonally stacked like ABAB. Since two different layers are used,

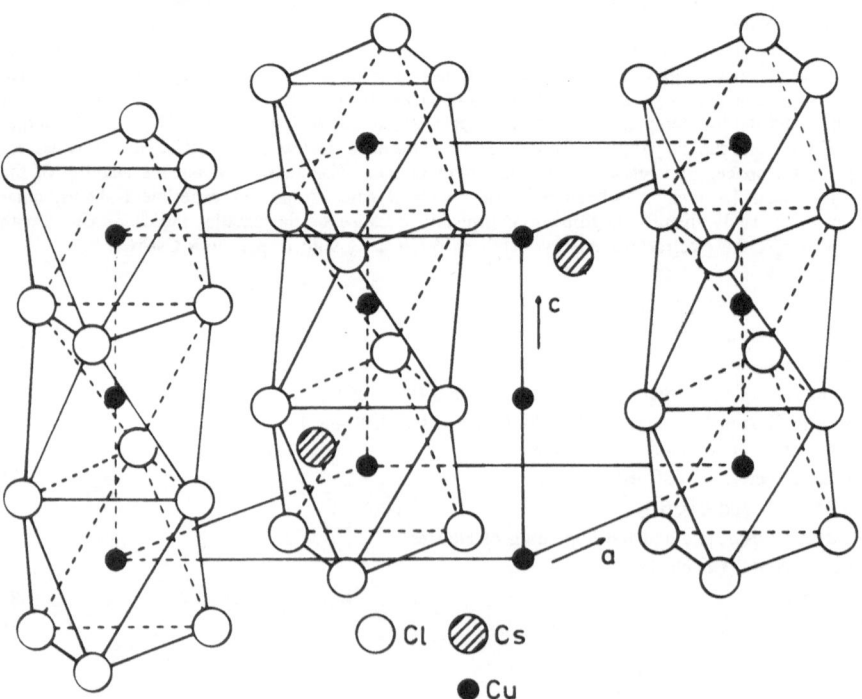

Fig. 1. Structure of α-$CsCuCl_3$ (P6$_3$/mmc). Aristotype of the hexagonal perovskites, which is also indicated by 2L or h-stacking. For reasons of clarity the column in the front at right has been omitted

one often refers to a 2L compound or to an *h*-stacking. In between the triangles of X-ions, the smaller B-ions are placed in octahedral holes. In these compounds the octahedra are face sharing and the B-ions form columns (Fig. 1). For structural problems these compounds have to be considered as three-dimensional, rather than one-dimensional, because of the strong correlations perpendicular to the c-axis.

Most of these compounds exhibit phase transitions, which can be described by a displacive and/or an order/disorder mechanism. At high temperature practically all of them have the structure of the undistorted hexagonal perovskites (e.g. $CsMgCl_3$). The phase transition at 473 K in $CsCuCl_3$ has been especially well studied [5]. Below this temperature an ideal (discrete) circular helix is formed. Structure determination by X-ray [6–8], measurements of the strain [9], neutron diffraction [10–14], and optical activity [15] have been done. The distortion below the phase transition can be described by a mode at $q = (0, 0, 2\pi/3c)$ [16, 17], so that the original c-axis is tripled. The transition appears to be of the first order type; this is proved experimentally by a small latent heat [5] and theoretically by studying the Landau–Lifshitz conditions for second order phase transitions [16, 17].

The fact that we have in $CsCuCl_3$ a helical compound, which can be described as a result of a displacive/order-disorder phase transition in an ideal hexagonal perovskite, is interesting, since the formation of the helix can be compared energetically with competing interactions of similarly built 2L compounds. There are other helical compounds (e.g. high-quartz, see [74]), where such an approach is not possible since in that case only reconstructive phase transitions can be imagined and then the gain in energy is much harder to estimate.

Also known are the compounds $KCuF_3$ [18, 19] and $KCrF_3$ where a cooperative $E \otimes \varepsilon$ JTE occurs in cubic perovskites. In cubic perovskites the sequence of close packed layers of the large ions (K^+ and F^-) is c-stacked (3L). The CuF_6-octahedra are corner-connected. The cooperative JTE is clearly evident in the CuF_2-planes where long and short axes are ordered in an antiferro-distortive way. This kind of ordering has also been found in K_2CuF_4, which certainly does not have the topology of a cubic perovskite. However, no transitions from the antiferro-distortive to a paradistortive phase have been found in these compounds. This difference with respect to the hexagonal perovskites was puzzling at first, since in the cubic perovskites only one anion bridges two JT-centres, whereas in hexagonal perovskites three anions are between two JT-ions, suggesting a weaker coupling for the cubic perovskites.

The reason for the different behaviour of face-sharing and corner-connected pairs of octahedra lies in the possibility of formulating local modes for the individual JT-octahedra. In the case of the hexagonal perovskites these local modes (which are of the E_g type) can be derived [20–22], and although these are not orthogonal between neighbouring centres their interaction can be overcome by raising the temperature without destroying the crystalline state. In the cubic perovskites, especially for the planes of CuF_2 and CrF_2, each F-ion has one coordinate, which affects both neighbouring centres in an equivalent way and no local coordinates can be found. Therefore, in the latter case the interaction between neighbouring centres is much stronger and cannot be overcome by raising the temperature without destroying the crystal.

In Sect. 2 the structures of β-CsCuCl$_3$ and β-RbCuCl$_3$ are described and compared. The notion of easy directions (Maaskant and Haije [26]) for shifts of close packed layers is worked out. A different view is given by not concentrating on the octahedron deformations, but by describing the changes of the Cl-triangles between two neighbouring Cu-ions. It is possible to show why two long Cu–Cl bonds never meet at the same anion. Also the observed strain-components are explained.

Since the replacement of Rb$^+$ by Cs$^+$ has drastic consequences, neutron diffraction measurements on acoustic phonon branches have been done on CsFeCl$_3$, as those on RbFeCl$_3$ had been measured before by Petitgrand et al. [23]. This is discussed in Sect. 3. The choice of non-JT compounds has the advantage that the complication of the cooperative JTE is avoided. An interpretation of these spectra is given. The phenomenon of structural resonance, which was first formulated by Heine and McConnell [24, 25] for incommensurate structures, and which proved to be essential for understanding the stability of helices in CsCuCl$_3$ [26] and TMCuCl$_3$ [27], is clearly seen in the LA- and TA- branch at the A-point in the first Brillouin zone (BZ) of the hexagonal cell of CsFeCl$_3$.

In Sect. 4 it is shown that the structure of CsCuBr$_3$, which has an *hc*-stacking (4L) of layers, can be derived from the hexagonal perovskites by instabilities at the A-point. Ting-i and Stucky [4] who determined the crystal structure and interpreted the magnetic measurements already concluded that a magnetic dimerization occurs. In our description there is a cooperative interference of JTE and antiferromagnetic superexchange which leads to the helical compound CsCuBr$_3$. Conceptually, the change in structure between CsCuCl$_3$ and CsCuBr$_3$ is a Spin–Peierls transition of a new kind.

The observed shear strain in RbCrCl$_3$ (Sect. 5) and the low temperature structure with space group C2 for CsCrCl$_3$, RbCrCl$_3$ and RbCuCl$_3$ show another type of structures competing with the helices found in CsCuCl$_3$. In the compounds with the C2 space group, pieces of the helices of CsCuCl$_3$ of both kinds occur alternatively. These can be called dl-compounds.

In Sect. 6 a general description of the structural helices, which can be derived from a more symmetrical crystal structure, is given. The question of what kind of distortion is needed for creating a helix will be examined. The conditions put forward by Heine and McConnell [24, 25] for structural resonance are discussed in detail.

Section 7 presents the discussion and the conclusions.

2 Comparison of the Structures of β-RbCuCl₃ and β-CsCuCl₃

In this section we concentrate on the difference between the low temperature structure of CsCuCl$_3$ and the so-called β-RbCuCl$_3$ structure. After the next section it will become understood why these structures differ from each other. This section, however, will deal with the easy directions for the shifts of planes and the dipolar moments in β-CsCuCl$_3$ and in β-RbCuCl$_3$. In addition the observed strain compounds can be rationalized.

The structural phase transition in CsCuCl$_3$ was discovered first by Kroese et al. [5] and experimental and theoretical studies of the phase transition associated with the realization of helically distorted chains have been carried out by many authors (Laiho et al. [28], Hirotsu [15], Fernández et al. [29], Khomskii [30], Lee [31], Vasudevan et al. [32], Tanaka et al. [33–36], Tazuke et al. [37, 38], Gesi and Osawa [39]). The phase transition was treated theoretically in terms of the phenomenological Landau–Lifshitz theory by Kroese and Maaskant [16] and Hirotsu [17]. The phase transition has also been studied by Khomskii [30] and Lee [31] in terms of the pseudo-spin phonon interaction and by Crama and Maaskant [40] in terms of the Potts model which is based on the theory of Höck et al. [41]. The structural helix in β-CsCuCl$_3$ has been explained in terms of structural resonance by Maaskant and Haije [26]. Schotte [12], Graf et al. [10, 11], and Schotte et al. [13] reported theoretical and experimental results of neutron scattering on the high temperature phase of CsCuCl$_3$.

Three phases of RbCuCl$_3$ have been observed by Crama [8, 42] and Harada [43, 44]. The theoretical description has been included in the mentioned studies of Crama and Maaskant [40], Harada [43, 44] and Tanaka et al. [33, 34]. The magnetic susceptibility measurements are reported by Tazuka et al. [71].

In Fig. 2 the space groups and the transition points are given for CsCuCl$_3$, RbCuCl$_3$, CsCrCl$_3$ and RbCrCl$_3$.

RbCrI$_3$ was found by Zandbergen and IJdo [78] to have a β-phase with space group C2/m at room temperature and the γ-phase (C2) at 1.2 K. This resembles the structures of RbCrCl$_3$, although the α-phase has not been found as far as we know. Above 150 K, CsCrI$_3$ [77] has space group P6$_3$/mmc. Below this temperature the β-form has the space group Pbcn [77]. Later this structure was also found for β-RbCuCl$_3$.

The site symmetry at the B-ions is strictly D$_{3d}$ in the space group P6$_3$/mmc of the aristotype structure. However, we have never found contradictions by assuming that the local octahedra are regular with symmetry O$_h$. The E $\otimes$ ε JTE of a regular octahedron is well known (e.g. [72]). In Fig. 3 the adiabatic potential surface is shown. It consists of a trough in the 2-dimensional space of the normal coordinates Q$_\theta$ and Q$_\varepsilon$ which belong to the ε-vibrations. With a difference of $2\pi/3$ the trough is warped (order of magnitude 50–200 cm^{-1}), denoting three equivalent more stable configurations. These stable configurations correspond in most Cu^{2+}- and Cr^{2+}-octahedra to an elongation of one of the axes of the octahedron and a shortening of the other two axes. Numerous studies, mainly

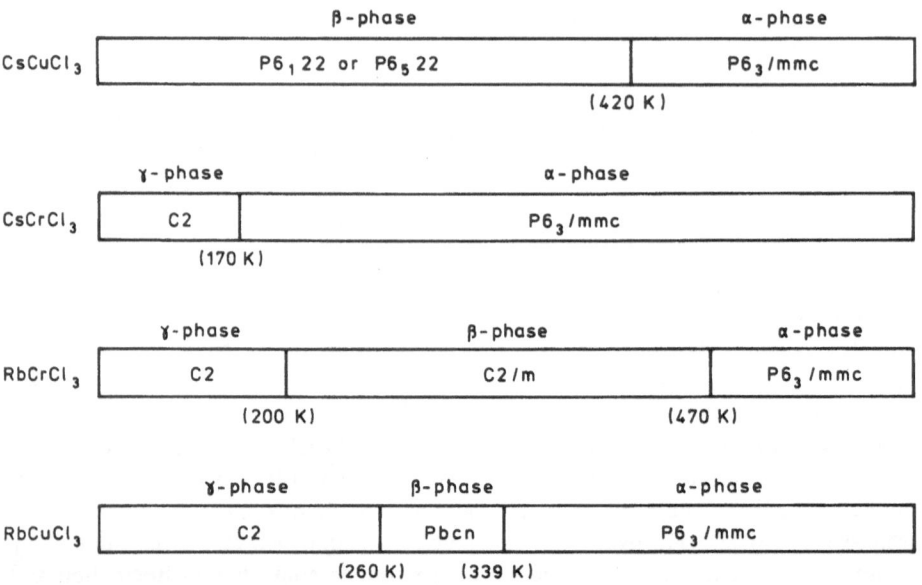

Fig. 2. Space groups and transition temperatures observed in CsCuCl₃, CsCrCl₃, RbCrCl₃ and RbCuCl₃

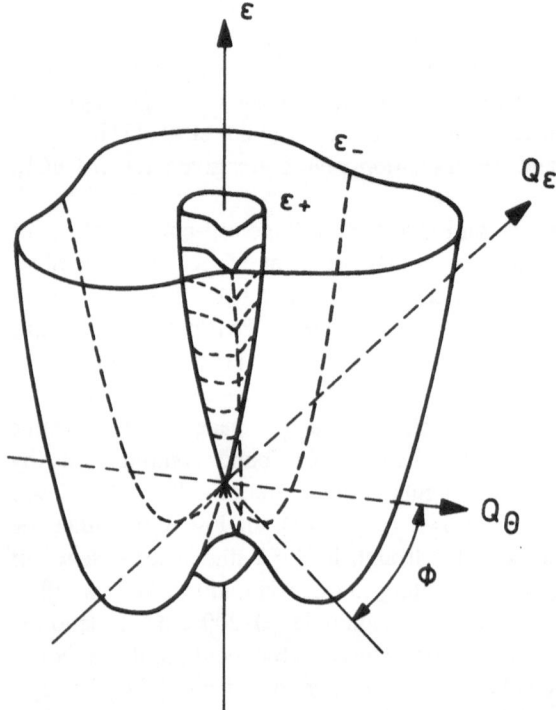

Fig. 3. Adiabatic potential energy surface for the $E \otimes \varepsilon$ Jahn-Teller effect in the two-dimensional space of the ε_g vibration modes

by X-ray diffraction analysis, of the geometry of $Cu^{2+}L_6$- and $Cr^{2+}L_6$-octahedra show these two oppositely oriented elongated B–L bonds and the remaining four shorter bonds (e.g. Reinen and Friebel [19]). In several cases, where a shortened octahedron has been found, either the results of the X-ray diffraction analysis were not sufficiently accurate, or an interpretation by a statistical average over two elongations has been possible, e.g. in β-RbCrCl₃ (Crama et al. [8, 46]), and (Garcia et al. [79]).

For the kind of structures that we describe in this paper, the elongation directions are simply the X-B-X directions, which are equivalent in the high symmetry structure. From X-ray diffraction it is easy to deduce how the arrangement of long axes is along the c-direction at lower temperature where the cooperative JTE is manifest. From Fig. 1 it is clear that there are two orientations of the octahedra. Figure 4 is a drawing of two of these face-sharing octahedra. At the same time, a notation of possible long axes is shown. Long axes are denoted by the ions they connect (D, E or F). But, due to the difference of orientation, those belonging to the lower octahedron are primed. Arrangements of long axes like D-D′, E-E′ and F-F′ are forbidden (see below). In β-CsCuCl₃ [1] the sequence of elongations along the c-axis is either DE′FD′EF′D or DF′ED′FE′D for the possible helices. In β-RbCuCl₃ the sequence is EF′EF′ or two other orientations. It is immediately seen that β-RbCuCl₃ retains the original c-axis (Fig. 1), whereas in β-CsCuCl₃ the c-axis is tripled. Furthermore, in the Cs-compound there is no enlargement of the cell in the a, b-direction, which means that all columns of face-coupled octahedra remain strictly parallel. For β-RbCuCl₃, however, the cell is enlarged in the a, b-direction and there are two kinds of orientation of strings.

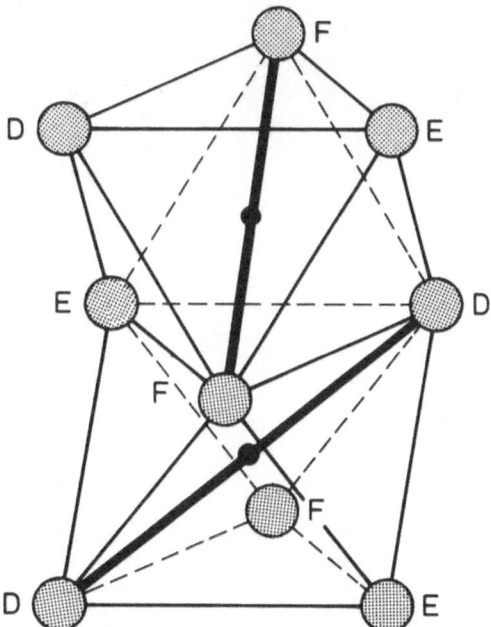

Fig. 4. Two face sharing octahedra. Elongated axes have thicker lines. This ordering is denoted as FD′, where the prime refers to octahedra oriented like the lower one

Apart from the question of how helical structures are built, there is the interesting question as to what the difference is between Cs^+ and Rb^+ that is decisive for the existence of the helix in the Cs-compound and not in the Rb-compound.

In order to discuss both structures we remark that the elongations are derived from the X-ion displacements. The X-ions of any octahedra belong to two neighbouring close packed layers. In the undistorted structure these form equilateral triangles. It is advantageous to express the relevant ε_g-modes in terms of shifts, tilts and deformations of these triangles. In the appendix these connections are described with the assumption that the Cu-centred octahedra are regular. The shifts, tilts and deformations show a different behaviour in the lattice of the hexagonal perovskites. Shifts of triangles make neighbouring ions in the same layer also shift and therefore lead to intralayer interactions. Tilts also tend to influence neighbouring layers and favour interlayer interactions. Deformations of X_3-triangles have only local influence. Another reason to focus on X_3-triangles is that each of them is influenced by a Cu-ion from above and from below and therefore distorts two Jahn–Teller octahedra.

From the appendix it is also seen that shifts of triangles are to be expected in the static JTE of one octahedron. It is necessary to study their behaviour since they tend to displace whole close-packed layers. As is well known, close-packed layers can have three positions in a stack – A, B or C. It is relatively easy to make a shift from ABA towards ACA. In the ideal structure this can be done in three equivalent directions, making angles of 120° with each other (Fig. 5).

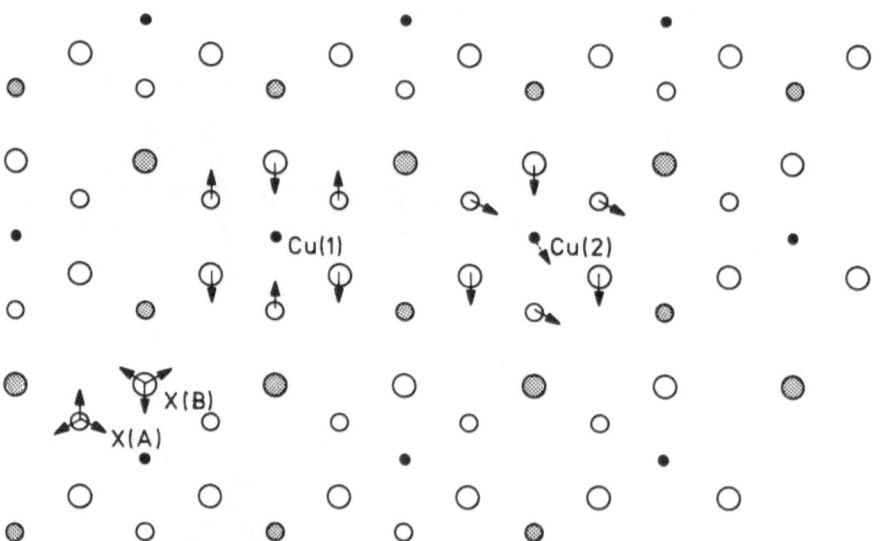

Fig. 5. Projection of two close-packed layers of large positive ions (*grey*) and negative ions (*open circles*). *The smaller black circles* represent Cu-ions. At X(A) and X(B) the directions of the easy shifts are indicated. The distortions around Cu(1) represent the shifts of the ions parallel to these planes for β-RbCuCl₃, and those for β-CsCuCl₃ are indicated at Cu (2). *The circles with the smaller radii* belong to the plane under those with the larger radii.

The direction of the arrows in this figure cannot be reversed, contrary to what one would expect in a harmonic model. A third order anharmonic term is present (Maaskant and Haije [26]) which takes into account that, by reversing the arrows, a layer sequence AAA is achieved which is energetically very unfavourable. Neighbouring planes have possible shifts which make angles of $\pm 60°$ or $180°$ as can be seen in Fig. 5 (X(A) and X(B)). In most cases these shifts are small so that the topology of the h-stacking is retained. But in $CsCuBr_3$ some shifts are so large that a change from ABABAB to ABACAB occurs and a hc-stacking arises (see below).

In β-$CsCuCl_3$ neighbouring AX_3-layers are shifted in directions making angles of $\pm 60°$ with each other. In β-$RbCuCl_3$ this angle is $180°$. As an immediate consequence the original inversion centre at the Cu-ions is retained in $RbCuCl_3$ and lost in $CsCuCl_3$ below the relevant transition points. This makes β-$CsCuCl_3$ a heli-electric compound whereas β-$RbCuCl_3$ has an antiferroelectric ordering. This is also illustrated in Fig. 5.

It can be explained here why elongation sequences such as DD' etc, where two long axes meet at the same anion, are rather improbable. The resultant shift would be twice that from one octahedron and would occur in the wrong direction favouring an AAA sequence of layers. In reality, two long axes never meet at one anion but are connected to different ions of an X_3-triangle. The resultant shift is only half as large as in the former case and its direction is reversed to an easy direction.

The tilts of the triangles in β-$RbCuCl_3$ and β-$CsCuCl_3$ are depicted in Fig. 6. These are easy to understand, since, from below the long axis the JTE pushes one Cl-ion up and from above another Cl-ion is pushed down. As remarked previously, β-$CsCuCl_3$ retains its hexagonal cell, while β-$RbCuCl_3$ becomes orthorhombic. Figure 7 shows the rectangular cell perpendicular to the c-axis. The arrangement of face-sharing strings is such that planes of Cl-ions parallel to the b, c-plane are shifted up and down alternately. This corresponds with a distortion belonging to the M-point in the first B.Z. (Fig. 10).

From Fig. 7 one also notices that the X_3-triangles are isosceles rather than equilateral and the unique angle in these triangles is larger than $60°$. This drawing is in proportion to the X-ray results and clearly demonstrates the distortion of the triangles. Suppose that in Fig. 4 the ions 1 and 2 are attached to long axes. The resultant deformation (see Appendix) is the sum of D_1 and D_2, where the normalized expressions, which differ by only a positive factor from the actual values are:

$$D_1 = \{-\tfrac{1}{2}X_3 - \tfrac{1}{2}\sqrt{3}Y_3 + X_2 - \tfrac{1}{2}X_1 + \tfrac{1}{2}\sqrt{3}Y_1\}/\sqrt{3} \tag{1}$$

$$D_2 = \{-\tfrac{1}{2}X_3 + \tfrac{1}{2}\sqrt{3}Y_3 - \tfrac{1}{2}X_2 - \tfrac{1}{2}\sqrt{3}Y_2 + X_1\}/\sqrt{3} \tag{2}$$

$$D_1 + D_2 = \{-X_3 + \tfrac{1}{2}X_2 - \tfrac{1}{2}\sqrt{3}Y_2 + \tfrac{1}{2}X_1 + \tfrac{1}{2}\sqrt{3}Y_1\}/\sqrt{3} . \tag{3}$$

For β-$CsCuCl_3$ the deformation of the triangles is similar as can be seen from Fig. 3A of Kroese and Maaskant [16].

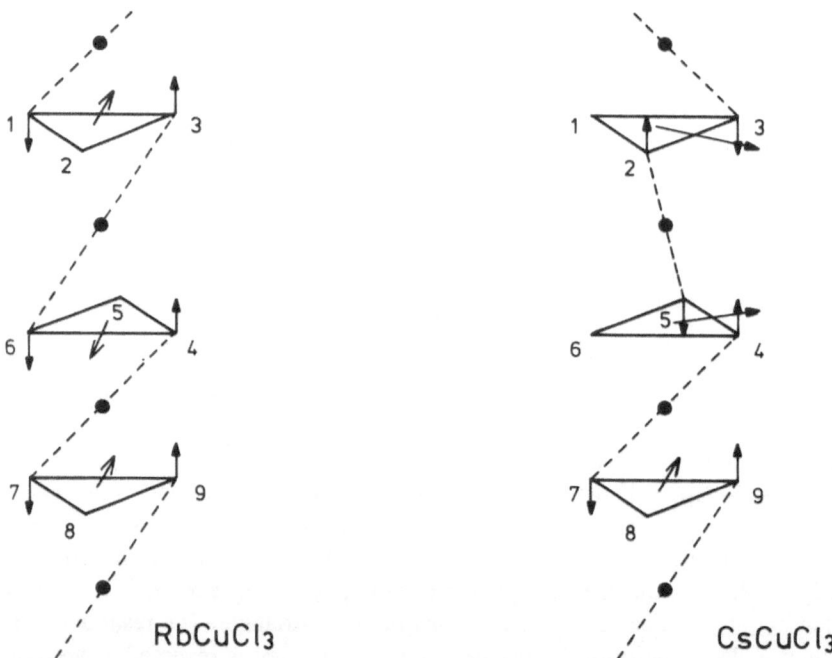

Fig. 6. Sketch of the z-directions of the Cl-ions in β-RbCuCl₃ (*left*) and β-CsCuCl₃ (*right*) along a column. The shifts of the triangles of the Cl-ions are indicated. These are attached to the points of gravity of these triangles. Note the helix in β-CsCuCl₃

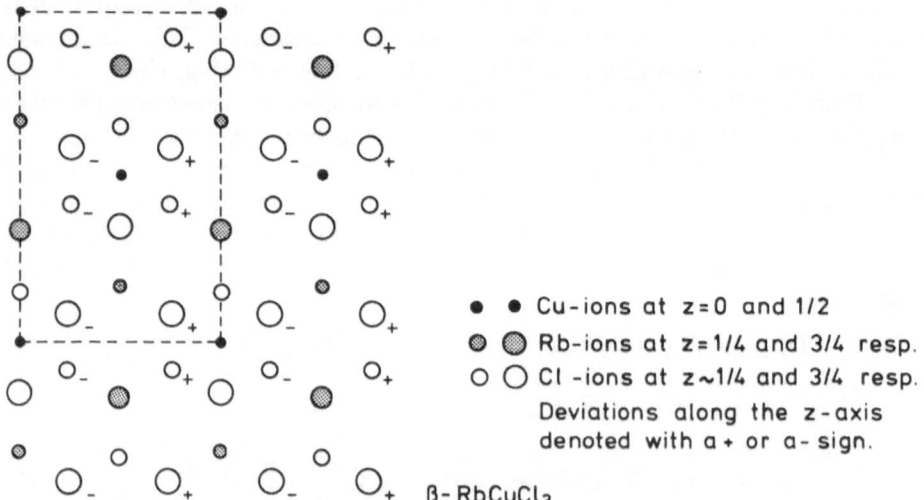

Fig. 7. Projection of the structure of β-RbCuCl₃ on the a, b-plane

Table 1. Crystal data of some $Cu^{2+}-$ and $Cr^{2+}-$ hexagonal perovskites. d is the distance between adjacent JT-ions. V, the volume of the cell, refers in all cases to $Z = 2$.

Compound	a(Å)	b(Å)	c(Å)	β(°)	d(Å)	V(Å³)	T(K)	Ref.
α-CsCuCl₃	7.229	–	6.149	90.00	3.075	278.22	470	7
α-CsCuCl₃	7.245	–	6.150	90.00	3.075	279.56	466	6
α-CsCuCl₃	7.212	–	6.141	90.00	3.071	276.57	430	7
β-CsCuCl₃	7.235	–	18.242	90.00	3.069	275.65	400	7
β-CsCuCl₃	7.216	–	18.178	90.00	3.062	273.24	293	1
α-RbCuCl₃	6.978	–	6.156	90.00	3.078	259.59	390	42
β-RbCuCl₃	11.929	6.971	6.164	90.00	3.084	254.26	293	42
γ-RbCuCl₃	11.932	6.844	12.244	91.93	3.061	249.83	4.2	42
α-RbCrCl₃	7.11	–	6.238	90.00	3.135	273.10	480	8
β-RbCrCl₃	12.224	7.04	6.25	93.34	3.125	268.84	295	46, 61
γ-RbCrCl₃	12.109	6.962	12.438	93.94	3.110	261.48	100	61
α-CsCrCl₃	7.257	–	6.238	90.00	3.119	284.50	295	8, 76
γ-CsCrCl₃	12.523	7.152	12.345	91.37	3.086	276.45	60	8, 61

The different angles between shifts of adjacent planes in β-CsCuCl₃ and in β-RbCuCl₃ also imply a different strain symmetry for these two compounds. In β-RbCuCl₃ an $(e_{xx} - e_{yy})$ strain of -0.0120 arises (Crama and Maaskant [40]). This strain component can be explained by noting that the elongation directions are not isotropically distributed in the direction of the basal plane. The formation of the helix in β-CsCuCl₃ also induces strain. This can be shown by the elastic measurements of the c_{33} coefficient (related to the c/a ratio) as a function of temperature (Lüthi [9]), which shows a continuous behaviour according to a second order transition. The shortening of the c-axis and the lengthening of the a-axis by lowering the temperature can be deduced from Table 1, where the cell parameters of CsCuCl₃ are given for different temperatures (Crama [8, 42]), see also Hirotsu [15]. Since there are shifts of layers perpendicular to the c-axis and distributed in six directions, the a- and b-axis will be enlarged. Nevertheless, the Cu–Cu distance (Table 1) remains approximately the same so that the c-axis is shortened. One has essentially the same effect when winding a circular helix from a piece of metallic wire of a certain length. The length along the axis of the helix will be less than the total length.

In the next section the origin of the difference in structure of β-CsCuCl₃ and β-RbCuCl₃ will become clearer.

3 Acoustic Phonon Branches of CsFeCl₃ Compared with RbFeCl₃ and the Phenomenon of Structural Resonance

As is described in the former section the cooperative JTE in CsCuCl₃ and RbCuCl₃ differ appreciably. Obviously this is caused by a different behaviour induced by the Cs- and the Rb-ion. Petitgrand et al. [23] measured the dispersion of some longitudinal as well as transversal acoustic phonon modes in RbFeCl₃ while we studied the acoustic phonon modes in CsFeCl₃ (Haije et al. [47]). Both compounds have undistorted hexagonal perovskite structures (Fig. 1). In order to observe the changes induced by the different alkali ions we thought it worthwhile not to include the complication of a cooperative JTE. Figure 8 reproduces the results for RbFeCl₃ and Fig. 9 gives our results of the phonon-dispersion curves. The longitudinal (LA) and transversal (TA) modes along the Γ–A, the Γ–K and the K–M in the first Brillouin zone (Fig. 10) are shown.

Obviously the phonon spectra of the two compounds are very different. In the case of RbFeCl₃, two features are remarkable. The transverse modes along the Γ–A line have an upward curvature, which is unusual and is, according to Petitgrand et al., due to the rigidity of the FeCl₃-chains. Secondly, the transverse mode along the Γ–K–M line softens as a function of temperature exactly at the K-point, although the frequencies of the whole line are relatively low. For RbFeCl₃ this softening is not so large that a phase transition occurs. But phase transitions related to an instability at the K-point have been observed in RbFeBr₃ [23, 48], TlFeBr₃ [49] and KNiCl₃ [50].

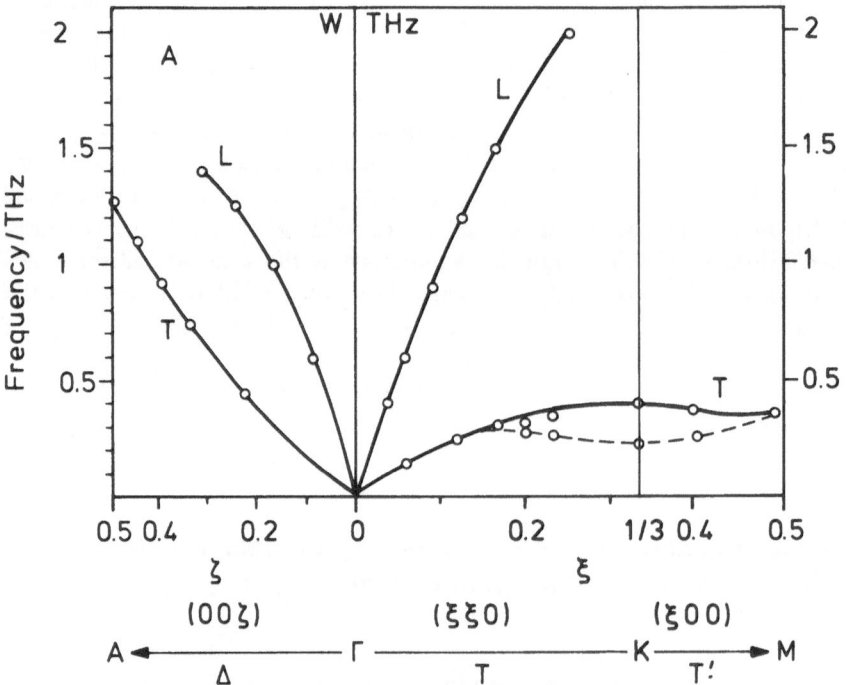

Fig. 8. Phonon dispersion curves for RbFeCl₃ (Petitgrand et al. [23]). *The dotted line* with a minimum at the K-point has been obtained by lowering the temperature

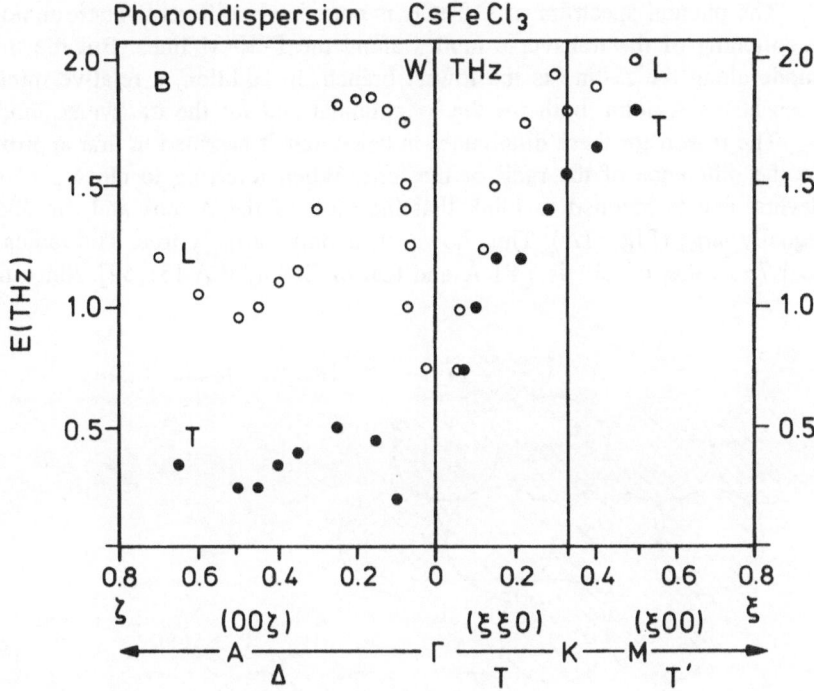

Fig. 9. Phonon dispersion curves for CsFeCl₃ (Haije et al. [47]). The dispersion along T and along T′ is uncertain due to twinning of the crystal. This did not affect the measurements along the Δ-line, however

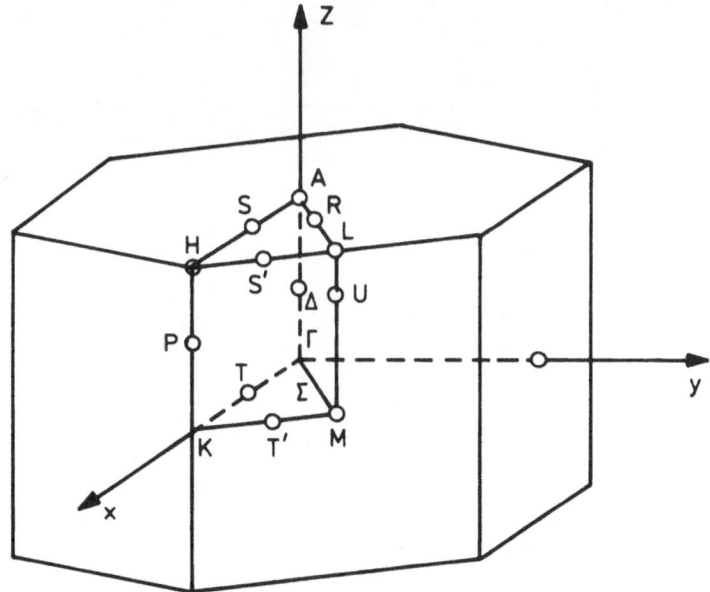

Fig. 10. First Brillouin zone of hexagonal lattices

The phonon spectrum of $CsFeCl_3$ is completely different. There is no sign of a softening of the transverse modes along the Γ–K–M lines. But the transverse mode along the Δ-line is the lowest branch. In addition, a relative minimum is seen at the A-point, both for the longitudinal and for the transverse mode.

The reason for these differences in behaviour is ascribed in first approximation to the difference of the radii of the ions. When referring to close packed AX_3-layers, one is inclined to think that the radii of the A-ions and the X-ions are equally large (Fig. 11A). This, however, is only partially true. The radius of Rb^+ is 1.73 Å, that of Cl^- is 1.81 Å and that of Cs^+ 1.88 Å [51, 52]. Since there are

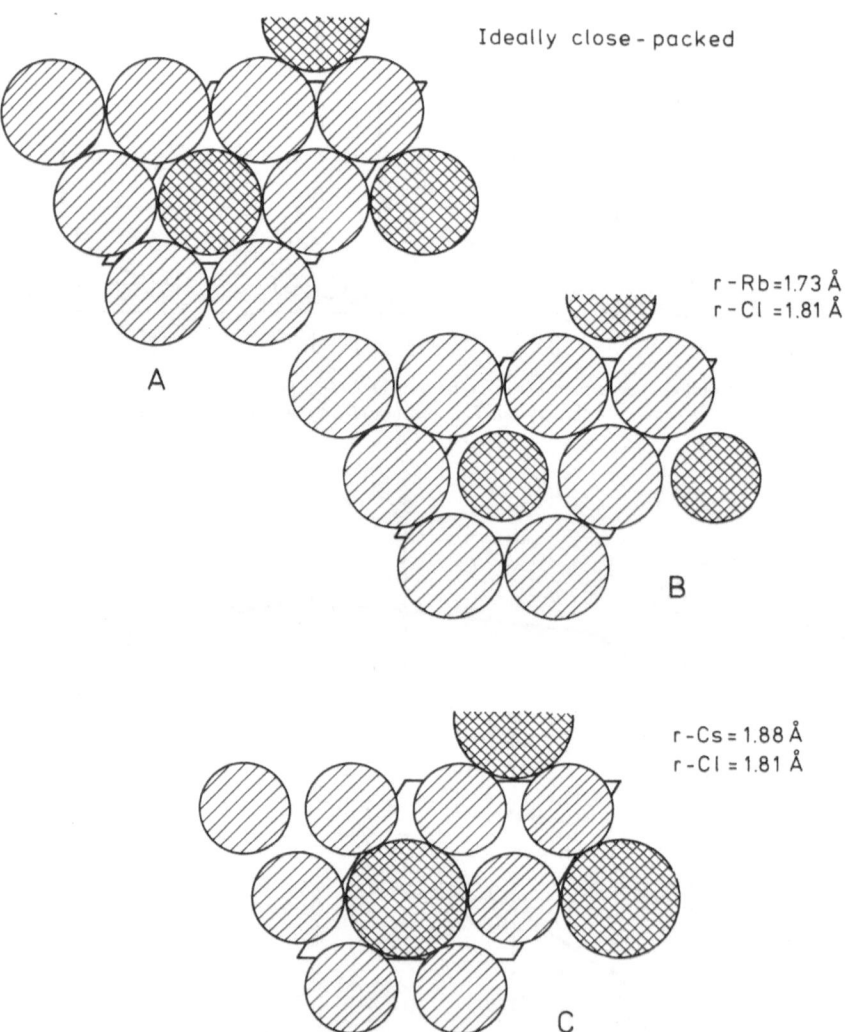

Fig. 11a–c. Close-packed layers of large positive ions and Cl-ions: a. Ideally close packed layers; b. the A-ion is somewhat smaller than the Cl-ions; c. the A-ion is somewhat larger than the Cl-ions

three times as many anions as Rb-ions in RbFeCl₃, the structure is still very rigid due to the exchange forces between the Cl-ions, so that these remain in place and form a rigid matrix in which there is a hole which is too large for a Rb-cation (Fig. 11B). In the case of CsFeCl₃, the anions are smaller than the Cs-ions and can therefore move more easily (Fig. 11C).

That the Cl-ions in CsFeCl₃ can move becomes clearer when the dips in the LA and TA branches are explained. These dips are the result of the phenomenon of structural resonance (Heine and McConnell [24, 25]. This phenomenon is also crucial for the building of helices [26, 27, 53], although it is of more general significance. In particular the dip in the LA branch at the A-point furnishes a one-dimensional example of this effect and will be explained.

Figure 12 shows how the Cs-ions vibrate in the z-direction with a wave length 2c, and how the Cl-triangles in between these Cs-ions move with a breathing mode parallel to the a,b-plane. A wave length corresponding to the A-point has been chosen on purpose, since the effect here is optimal as will be demonstrated by a simplified model. Suppose the z-displacements of the Cs-ions are denoted by m_i, where the index i indicates the number of the Cs-ion. Likewise, Q_j represents the breathing coordinate from the Cl-triangle numbered j. Note that the symmetry of the two modes is different. m_i is antisymmetric for the plane parallel to the a,b-plane and through i, while Q_j is symmetric for reflection in such a plane through j. One would not expect interaction between two modes which behave differently with respect to these planes. This, however, is only true for a wave at $q = 0$. Heine and McConnell [24, 25] showed that this is generally not the case when $q \neq 0$. Energy lowering can occur through interaction of these modes,

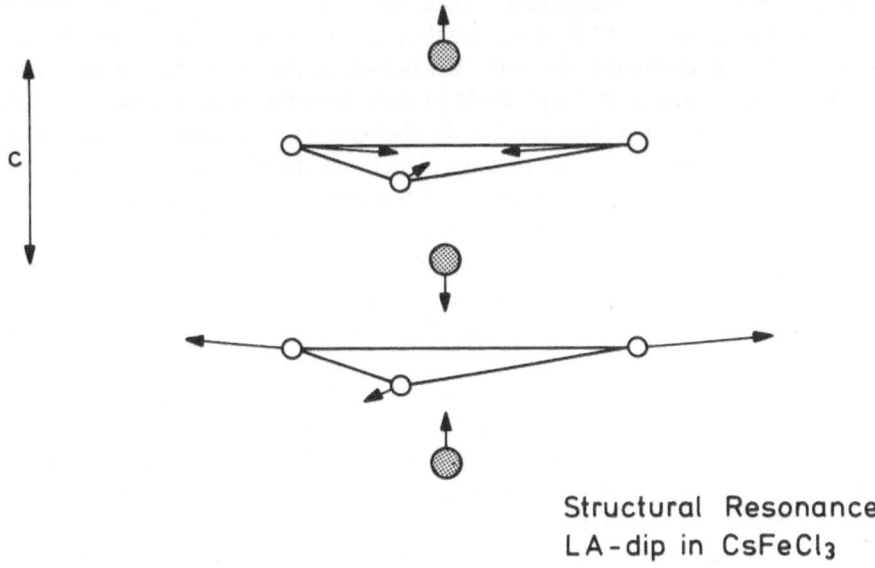

Structural Resonance
LA-dip in CsFeCl₃

Fig. 12. Structural resonance in the LA branch at the A-point in CsFeCl₃. The interacting breathing mode of the Cl-triangle and the Cs z-dispacements are shown

which has been named 'Structural Resonance'. This can be shown in a harmonic approximation. In real space the following form of the bilinear interactions can be written

$$H = -\frac{1}{2}\sum_{ij}|Q_i m_i| \begin{vmatrix} F_{ij} & S_{ij} \\ T_{ij} & G_{ij} \end{vmatrix} \begin{vmatrix} Q_j \\ m_j \end{vmatrix}.$$ (4)

F_{ij}, G_{ij}, S_{ij} and T_{ij} are interaction constants between neighbours, next nearest neighbours etc. The factor $1/2$ prevents double counting. When working these terms out it is seen that $T_{ij} = S_{ji}$. The interaction S_{ij} should not disappear, e.g. on symmetry reasons, since then there is never interaction between the Qs and the ms. The expression for the interaction Hamiltonian is now Fourier transformed:

$$H = -\frac{1}{2}\sum_{q}|Q_{q^*} \ m_q| \begin{vmatrix} F(q) & S(q) \\ S(q)^* & G(q) \end{vmatrix} \begin{vmatrix} Q_q \\ m_q \end{vmatrix}.$$ (5)

Here

$$F(q) = F(0) + F_1\cos(q_z c/2) + F_2\cos(q_z c) + \ldots$$
$$G(q) = G(0) + G_1\cos(q_z c/2) + G_2\cos(q_z c) + \ldots .$$ (6)
$$S(q) = iS_1\sin(q_z c/2) + iS_2\sin(q_z c) + \ldots$$

The wave vector is restricted to plane waves along the c-axis. The constants $F(0)$ and $G(0)$ refer to interactions from neighbours in the plane. For S there are no such interactions. The indices 1 and 2 refer to interactions with neighbours and next nearest neighbours respectively along the c-axis, for which the distances are $c/2$ and c respectively. The interaction $S(q)$ between the Q- and the m-mode is zero for $q_z = 0$ as expected. When the interaction $S(q)$ can be restricted to nearest neighbours, $S(q)$ is maximal when $q_z = \pi/c$, that is exactly at the A-point. This explains the dip in the LA branch at that position. The last term in Eq. (6), when sufficiently large, tends to shift the minimum to another q_z-value.

The dip in the two TA-branches is also due to structural resonance, but is a little more involved, as the phonon modes are doubly degenerate. In Fig. 13 these movements are sketched, again at the A-point. The displacements of the Cs-ions are now in the x- and y-directions and are symmetric for reflection in the plane parallel to the a,b-plane through the Cs-ions. These displacements interact with tilts of the X_3-triangles in corresponding directions. Note that these tilts are antisymmetric for reflection in the planes through their equilibrium positions perpendicular to the c-axis. Again, when the interaction between next nearest neighbour planes can be neglected, the maximal interaction and the dip are at the A-point. At the Γ-point ($q = 0$) the interaction is zero.

In conclusion, we have shown that, through the difference in radii of the Rb- and the Cs-ions, quite different movements of the Cl-ions occur. It remains to be deduced what this means for the cooperative JTE in $RbCuCl_3$ and $CsCuCl_3$.

The influence of the A-ion for the cooperative JTE has been neglected so far in the literature. But, through its anion neighbours, an A-ion is connected with three columns of face sharing octahedra and therefore, in this case, with

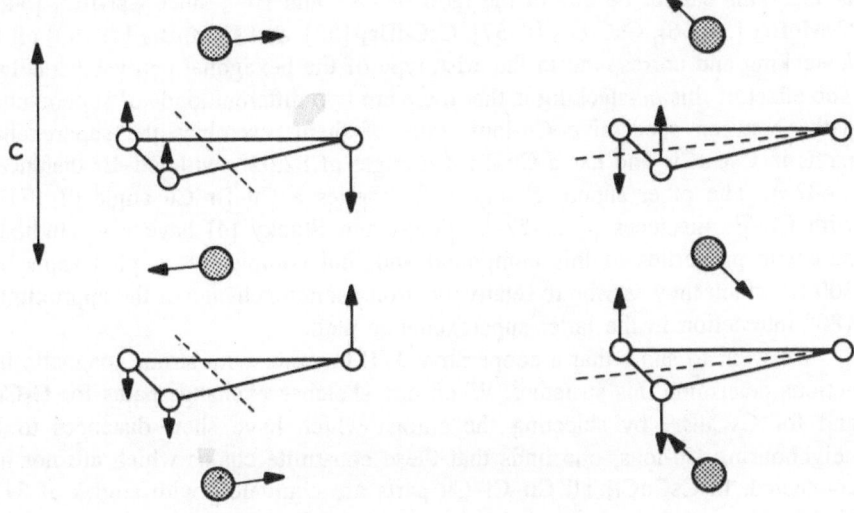

Structural Resonance
TA -dip in CsFeCl₃

Fig. 13. Structural resonance in the two-fold degenerate TA-branch at the A-point for CsFeCl₃. Tilts of the Cl-triangles cooperate with the Cs x- and y-displacements. Modes of this branch together with the cooperative JTE are responsible for the helices in β-CsCuCl₃, CsCuBr₃, TMCuCl₃ and the C2-structures

six Cu-ions. Considering both the Rb- and the Cs-compounds, a compromise is found between the JTE and the requirements of the A-ions. The final structure is reached when the cooperative JTE effect is combined with the lowest acoustic branch. In the case of RbCuCl₃ this occurs at the M-point and not at the K-point, but the energy difference is not large as can be seen in Fig. 8. For β-CsCuCl₃ the phase transition is at $q_z = 2\pi/3c$ and not at the A-point, the reason being that at this value there exists a third order invariant [16], connected to the fact that each individual octahedron wants to occupy a well in its trough (see Fig. 3). Indeed, advantage is taken of both cases of low lying acoustic modes, but a compromise had to be found in order to satisfy the requirements of both the JT-ion coupling and the effect of the A-ion.

4 On the Structure of CsCuBr₃

This compound has been investigated by Ting-i and Stucky [4]. It has a so-called *hc*-stacking of layers. This means that the aristotype has a layer sequence of the Cs- and the Br-ions corresponding to ABAC, repeating itself after four layers. This cannot be due to the radii of Cs^+ and Br^-, since $CsNiBr_3$ [54, 55], $CsMgBr_3$ [55, 56], $CsCrBr_3$ [8, 57], $CsCdBr_3$ [55], and $CsMnBr_3$ [58–60] all have h-stacking and correspond to the aristotype of the hexagonal perovskites (Fig. 1). The effect of this hc-stacking is that there are two different kinds of superexchange paths between successive Cu-ions. One of them resembles the superexchange paths in $CsCuCl_3$ and has a Cu–Br–Cu angle of 82.02°, with Cu–Br distances of 2.482 Å. The other superexchange path implies a Cu–Br–Cu angle of 171.46°, with Cu–Br distances of 2.522 Å. Ting-i and Stucky [4] have also studied the magnetic properties of this compound and find completely coupled spins up to 300 K, which they ascribe to relatively strong superexchange of the approximately 180°-interaction in the latter superexchange path.

We want to show that a cooperative JTE together with strong magnetic inter-actions determine this structure. When one sketches exchange paths for $CsCuCl_3$ and for $CsCuBr_3$ by selecting the anions which have short distances to both neighbouring Cu-ions, one finds that these constitute chains which are not inter-connected. In $CsCuCl_3$ all Cu–Cl–Cu parts are equivalent with angles of 81.12° [1], forming a relatively weak interaction. In $CsCuBr_3$ these chains have dimer-ized by opening every second Cu–Br–Cu angle to 171.46°, so that each Cu-ion contributes to one strong and one weak superexchange interaction. If one compares ideal h- and hc-structures, it is difficult to decide by what shift the hc-structure is formed from the h-structure, since there are three different but equivalent directions (making angles of 120°) to shift, say, an A-layer towards the C-layer position. However, in $CsCuBr_3$ one can indicate by following the superexchange paths the Cu-ions which left the original column in the hexa-gonal perovskite structure. It is therefore possible to imagine a displacive phase transition between the h- and the hc-structure. This transition occurs with a q-vector at the A-point in the first Brillouin zone, since the c-axis is doubled with respect to the original hexagonal perovskite. General expressions for the shifts (m_x, m_y) and the quadrupole moments (Q_{xz}, Q_{yz}) of the Cu-ions are (Maaskant and Haije [26]):

$$\mathbf{m_q} = (2N)^{-1/2}\sum_n (-1)^n \mathbf{m_n} \exp(i\mathbf{q} \cdot \mathbf{R_n}) \tag{7}$$

$$\mathbf{Q_q} = (2N)^{-1/2}\sum_n (-1)^n \mathbf{Q_n} \exp(i\mathbf{q} \cdot \mathbf{R_n}) . \tag{8}$$

Here $\mathbf{q} \cdot \mathbf{R_n} = q_z z_n = \pi z_n/c$ and N is the number of unit cells.

One can think of the $h \rightarrow hc$ transition as starting by a wave, which denotes the relative shift m_x of the layers and the Cu-ions as given in Fig. 14 by

$$m_{xq} = -(2N)^{-1/2}\sum_n (-1)^n m_{n0x} \cos(\pi z_n/c) \tag{9}$$

where, e.g., the Cu-ions are numbered from the top downwards by 1, 2, etc. The m_{n0x} are equal amplitudes for each Cu-ion and point in the positive x-direction and c refers to the c-axis of the original h-cell. In Fig. 14 the phases correspond to $z_n = (2n - 1)c/4$. One might argue that in one half of the cosine wave the

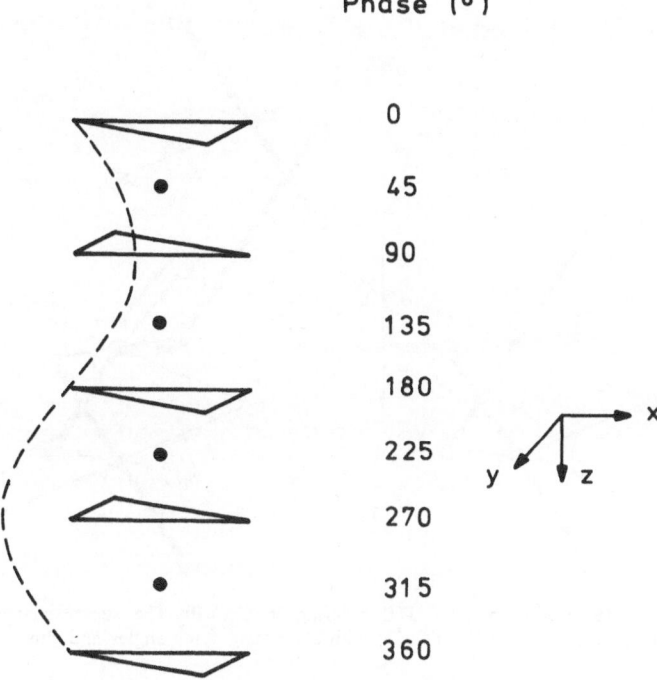

Phase (°)

0

45

90

135

180

225

270

315

360

Fig. 14. Shift of layers in an easy direction such that *h*-stacking is converted into *hc*-stacking. This shift belongs to a wave at the A-point (q=(0, 0, π/c) with respect to the *h*-lattice

movement of the layers is opposite to an easy direction (Sect. 2). However, this can be remedied by adding a constant shift to the crystal, which does not require energy and which can have such a value that everywhere the displacement is in an easy direction.

It would be a mistake to draw long axes from the JTE in Fig. 14, since the observed arrangement is fully adapted to the *hc*-structure. In Fig. 15 a slightly idealized picture of the observed structure is given. It consists of layers of equally oriented dimers of face sharing $CuCl_6$-octahedra. Each dimer is connected to three similar dimers in the layer above and below. A 2-fold axis through Br(b) practically perpendicular to the plane of the paper connects these dimers. There is also a 2_1-axis, parallel to z, which passes close to Br(b) and which connects these two dimers. Each dimer also has a 2-fold axis through Br(a) as indicated in one case. A third dimer in the lower slab, which is also connected to the upper dimer and which contains Cu-ion nr. 7, is not drawn for reasons of clarity. Br(a) and Br(b) correspond to Br(1) and Br(3) of the structure determination (Ting-i and Stucky [4]).

Figure 15 has been drawn in order to show the superexchange path (Br(b)–Cu(4)–Br(a)–Cu(3)–Br(b)–Cu(1)–) and the cooperative JT ordering. The + and the − sign indicate whether Br-ions are in front of or behind the Cu-ion of the

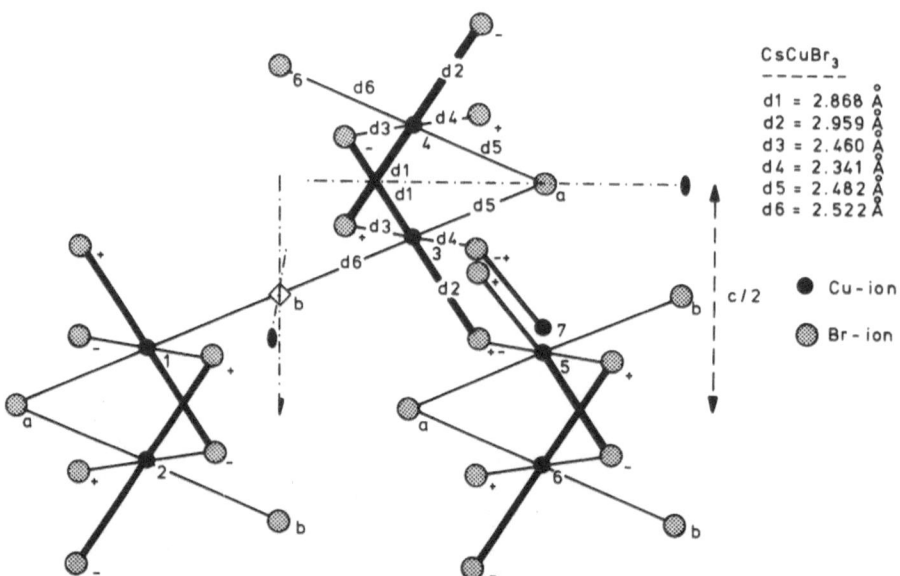

Fig. 15. The cooperative JTE ordering in CsCuBr₃.The superexchange path -Cl(b)-Cu(4)-Cl(a)-
Cu(3)-Cl(b)-Cu(1)-Cl(a)-Cu(2) with alternating acute angles and almost straight lines is indicated

octahedron they belong to. The Br-ion connecting Cu(3) and Cu(5) is in front
of Cu(3), but behind Cu(5). The long bonds, due to the JTE, are denoted by a
thicker line. Once the exchange-path is chosen, there are two possible orientations
for the JTE in a dimer. If one of them is chosen, the ordering in successive layers
is also fixed, when it is assumed that all dimers in the x, y direction are equally
oriented and that an arrangement with two long axes attached to one Br-ion is
not allowed.

The displacements of the Br-ions in each octahedron can be described in a first
approximation by quadrupoles situated on the Cu-ions and hence by equations
similar to Eq. 8. In particular the quadrupole normalized symmetry modes are:

$$Q_{qxz} = -(2N)^{-1/2} \sum_{n}^{2N} (-1)^n Q_{n0xz} \sin(\pi z_n/c) \qquad (10)$$

$$Q_{qyz} = -(2N)^{-1/2} \sum_{n}^{2N} (-1)^n Q_{n0yz} \cos(\pi z_n/c) . \qquad (11)$$

Here Q_{n0xz} and Q_{n0yz} are unit quadrupoles attached to Cu(n), which are all oriented in the same way and where $n = 1\ldots 2N$ assuming N unit cells in the
c-direction.

In the harmonic theory the amplitudes Q_{0xz} and Q_{0yz} are related. However,
the amplitude of the displacement mode is so large that anharmonicity is expected. The quadrupole amplitudes can adjust themselves independently, so that
in each octahedron the long axes can be placed in accordance with the local

minima in the JT trough. Nevertheless the occurrence of these two quadrupole amplitudes reminds us of a helical structure. The space group of this structure $C222_1$ (D_2^5) is enantiomorphic. Changing the sign of the expression of Q_{qyz} (Eq. 11), transforms the structure into its enantiomorphic partner. This is similar for the helical compound $CsCuCl_3$.

The structure is orthorhombic with axes a = 12.776, b = 7.666 and c = 12.653 Å. A relative ($e_{xx} - e_{yy}$)-strain of -0.0378 can be deduced, which differs appreciably from the values found in e.g. β-$RbCuCl_3$ and γ-$CsCrCl_3$ which are -0.0120 and 0.0108 respectively [40].

5 On $RbCrCl_3$ and $CsCrCl_3$

Preparations, crystal structure determinations and magnetic susceptibility measurements on $RbCrCl_3$ and $CsCrCl_3$ have been done by Crama [7, 8], Crama et al. [46, 61], Crama and Zandbergen [76], Leech and Machin [62, 63], and Larkworthy and Trigg [64], neutron scattering studies by Graf et al. [10, 11], Schotte [12], Schotte et al. [13], and Pyka et al. [14]. Theoretical descriptions were performed by Crama and Maaskant [40], Tanaka et al. [33, 34], and Perez-Mato et al. [65, 66]. We would also mention the structure determinations and magnetic measurements on $RbCrI_3$ and $CsCrI_3$ by Zandbergen and IJdo [77, 78].

It is our intention to show that the magnetic properties of the columns of face-sharing octahedra containing Cr also influence the interaction energies between the octahedra along a column of $RbCrCl_3$ and $CsCrCl_3$. We present here a new interpretation of the magnetic properties of these compounds. Cr^{2+} (d^4) in these compounds is high spin. The spin forbidden transitions from the ^5E ground state to the triplet states lie at $\gtrsim 6500$ cm^{-1} for $RbCrCl_3$ and $CsCrCl_3$ [85]. There are two ways of magnetic interaction between two Cr-ions in neighbouring octahedra: by direct exchange from t_{2g}-orbitals, one from each ion and both directed along the c-axis and by interaction of the e_g-orbitals through the three Cl-ions of the shared face. In our view the experiments show that the magnetic interaction of the e_g-electrons is ferromagnetic. This follows from the fact that the corresponding solids $RbVCl_3$ and $CsVCl_3$, which have no e_g-electrons, are more strongly antiferromagnetic than $RbCrCl_3$ and $CsCrCl_3$. The J/k ratios for $RbVCl_3$, $CsVCl_3$, $RbCrCl_3$ and $CsCrCl_3$ are, respectively, -123 K [68], -115 K [68], -24 K [8, 63] and -20 K [8, 63]. Similar results are found for $CsVI_3$ and $CsCrI_3$ [69], which have, respectively, J/k = $(-54 \ldots - 67)$ K [68] and -14 K [78].

The wave function for two electrons (in Cr^{2+} one e_g-electron from each octahedron of a neighbouring pair) or for two e_g-holes (in Cu^{2+} one from each octahedron) is given by (e.g. Kahn [67]).

$$\Psi = (1/\sqrt{2})[\chi_A(1)\chi_B(2) \pm \chi_B(1)\chi_A(2)] . \tag{12}$$

Since the total wave function has to be antisymmetric, the $+$ sign refers to a singlet and the $-$ sign to a triplet state. The repulsion between the two electrons will give rise to an energy:

$$E' = (1/2) \int [\chi_A(1)\chi_B(2) \pm \chi_B(1)\chi_A(2)]^2/r_{12} \, d\tau = J \pm K . \qquad (13)$$

Here the Coulomb integral J is

$$J = \int |\chi_A(1)|^2 |\chi_B(2)|^2/r_{12} \, d\tau = \int |\chi_A(2)|^2 |\chi_B(1)|^2/r_{12} \, d\tau \qquad (14)$$

and the exchange integral K is

$$K = \int \chi_A^*(1)\chi_B(1)\chi_A(2)\chi_B^*(2)/r_{12} \, d\tau . \qquad (15)$$

Both integrals are positive, because of the Jahn–Teller effect, there being an additional degree of freedom. According to Abragam and Bleaney [70] the electronic wave function of the lowest energy in a regular octahedron, assuming the linear coupling constant to be positive, (see also Ham [83, 84]) is

$$\Psi_- = d_\theta \cos(\phi/2) - d_\epsilon \sin(\phi/2) . \qquad (16)$$

Here d_θ and d_ϵ are, according to molecular orbital theory (e.g. [80]),

$$d_\theta = d(z^2) \cdot \cos \alpha + \sqrt{(1/12)} \, (2z_3 - z_1 - z_2 + 2z_6 - z_4 - z_5) \cdot \sin \alpha \qquad (17)$$

$$d_\epsilon = d(x^2 - y^2)\cos \alpha + (1/2) \, (z_1 - z_2 + z_5 - z_6) \cdot \sin \alpha . \qquad (18)$$

The z_i are σ-type orbitals directed along the bond directions. α describes the mixing between ligand and central ion wave functions. The e_g orbitals of neighbouring octahedra meet at the common ligands. We assume two regular face sharing octahedra. ϕ_1 and ϕ_2 are the orientation angles for these octahedra. It is then possible to derive, that

$$J = C_1\{12 + 6\cos(\phi_1 - \phi_2)\} \qquad (19)$$
$$K = C_2\{12 + 6\cos(\phi_1 - \phi_2)\} . \qquad (20)$$

The constants C_1 and C_2 have to be calculated numerically ($C_1 > C_2 > 0$). This follows since C_2 is a measure of the self energy of the overlap density only, whereas C_1 is the repulsion between the total charge density of the e_g-wave functions on one and the other side. This means that E' from Eq. (13) gives

$$E' = (C_1 \pm C_2) \, \{12 + 6\cos(\phi_1 - \phi_2)\} . \qquad (21)$$

The lowest interaction energy occurs for the minus sign and for $(\phi_1 - \phi_2) = \pi$:

$$E' = 6(C_1 - C_2) . \qquad (22)$$

This means that the interaction of the spins connected to the e_g-wave functions is ferromagnetic and that if, e.g., $\chi_A = d_\theta$, $\chi_B = \pm d_\varepsilon$. Therefore, the e_g-orbitals of neighbouring octahedra are preferably orthogonal, which is also in accordance with Hund's first rule: "Other things being equal, the state of the highest multiplicity will be the most stable" (formulation by Kauzmann [81]).

This ferromagnetic interaction is demonstrated by $CsCuCl_3$, where, from magnetic susceptibility measurements, Tazuke et al. [37] found that, along the columns, $J/k = 24$ K. Also Adachi et al. found that the interaction along the columns is ferromagnetic [86]. Indeed, since $(\phi_1 - \phi_2) = \pm 2\pi/3$, Eq. (21) gives

$$E' = 9(C_1 \pm C_2) \tag{23}$$

which leads to a ferromagnetic interaction.

In Sect. 2 we gave reasons why two long axes never meet on a ligand. Here we have an additional argument, since, for this to occur, $\phi_1 = \phi_2$. This gives as lowest energy

$$E' = 18(C_1 - C_2) \tag{24}$$

which differs from Eq. (22) by the positive quantity

$$12(C_1 - C_2).$$

In $RbCrCl_3$ and $CsCrCl_3$, the antiferromagnetic direct exchange between nearest neighbouring Cr-ions due to t_{2g}-orbitals is stronger. Also the Cr-ions remain in the $S = 2$ state. Therefore the spins of the electrons of the overlapping e_g-orbitals are opposite with respect to each other and the resulting J-values are less than the ones without e_g-electrons, as observed.

In the first discussion on the structure of β-$RbCrCl_3$ Crama and Maaskant [8, 40] indicated that the stabilization of this structure is due to entropy, since from X-ray experiments they could deduce that two elongations were involved. We still adhere to this picture, but one realizes, in view of Eq. (22), that the barrier between the two elongation directions in $RbCrCl_3$ and $CsCrCl_3$ can very well be lowered as a result of this ferromagnetic interaction. In view of the fact that $(\phi_1 - \phi_2) = \pm 2\pi/3$ or π we expect E' to equal $9(C_1 + C_2)$ or $6(C_1 + C_2)$ respectively.

The lowering of this barrier in these compounds due to superexchange is not strong enough to overcome the wells in the Jahn–Teller trough completely. This follows from the structures of γ-$RbCrCl_3$ and γ-$CsCrCl_3$ (e.g. [40]) where every Cr-ion is in one of the three wells which also occur in isolated octahedra.

When one compares the structures of β-$CsCuCl_3$ and $CsCrCl_3$, which has no β-phase, then the only difference is in the Cu^{2+}- and Cr^{2+}-ion. Lüthi [9] observed in crystals of $CsCuCl_3$ a softening of the c_{44} elastic constant. By extrapolating this value to zero, one can estimate a transition point of 324 K [17] to the β-$RbCrCl_3$ structure for $CsCuCl_3$. This means that in $CsCuCl_3$ the helical structure is about 100 K more stable than the one of α-$CsCrCl_3$ or β-$RbCrCl_3$.

The reason for the β-phase being absent in CsCrCl$_3$ is most probably a combination of the effect of RbCrCl$_3$ and the fact that the Cl-ions are rather loose, as we saw in Sect. 3. At very low temperatures an ordering with space group C2 occurs (γ-CsCrCl$_3$, γ-RbCrCl$_3$, γ-RbCuCl$_3$). This occurs, e.g. in β-RbCrCl$_3$, by an additional mode of type A$_3$ which gives the ordering of the long axes:

zxzyzxzy .

This A$_3$-mode at the A-point in the first Brillouin zone is compatible with Δ$_5$, which is responsible for the helix in CsCuCl$_3$. We now translate the sequence of long axes by replacing the symbol for each octahedron with h when its neighbours are equal and with c or d when its neighbours are unequal. We use c when xyz are in alphabetical order or a cyclic change of that, d being used for the other permutations of xyz. The result is

chdhchdh .

When one compares this with β-CsCuCl$_3$, which can be expressed as

cccccc or dddddd

for either the P6$_1$22 or the P6$_5$22 space groups, we can describe the C2 structure as a racemic compound of an equal number of right- and left-handed helices. Crystals of dl-compounds have been studied for the first time by Pasteur (1848) [82] and Roozeboom (1899) [82] on dl-tartaric acid.

We have calculated the shifts of the Cr-ions in γ-CsCrCl$_3$ from the averaged position. These are for Cr(1a), Cr(2), Cr(1b) and Cr(2'), respectively, 0.109, 0.120, 0.184 and 0.120 Å and should be compared with the shift of Cu in β-CsCuCl$_3$ which is 0.445 Å. This means that the h-succession plays an important role in the structure of the Cr-compounds. On the other hand the helix in CsCuCl$_3$ has a larger electrostatic energy. Possibly the difference in structure type for the distorted Cr-compounds and β-CsCuCl$_3$ is caused by the difference in size of the transition metal ions. Shannon and Prewitt [52] gave radii for Cr^{2+} and Cu^{2+} in 6-coordination of respectively 0.82 and 0.73 Å. We find for the distorted Cr- and Cu-compounds as determined by Crama [8], on the average 0.71 and 0.65 Å respectively. It is also possible that the difference in radii which are estimated for Cr^{2+} and Cu^{2+} to be respectively 0.82 and 0.73 Å in 6-coordination [52, 53] is important in the choice of the different types of structures. (We find, as an average, for the different β-RbCrCl$_3$, γ-RbCrCl$_3$ and γ-CsCrCl$_3$, as determined by Crama [8], 0.71 Å for the radius of Cr^{2+} and from the structures of β-RbCuCl$_3$, γ-RbCuCl$_3$ and β-CsCuCl$_3$ 0.65 Å for the radius of Cu^{2+}.)

Another question arising is why the shear strains in γ-RbCrCl$_3$ and γ-CsCrCl$_3$ have a different sign. In Fig. 16 the two structures have been displayed in proportion to the measured crystal data. The black dots represent the Cr-ions. For two of them the long axes (previously denoted as z) are indicated. These are Cr(1a) and Cr(1b). When local coordinate axes are assumed these two elongations can both be expressed as Q(z^2). For the other two octahedra we denoted

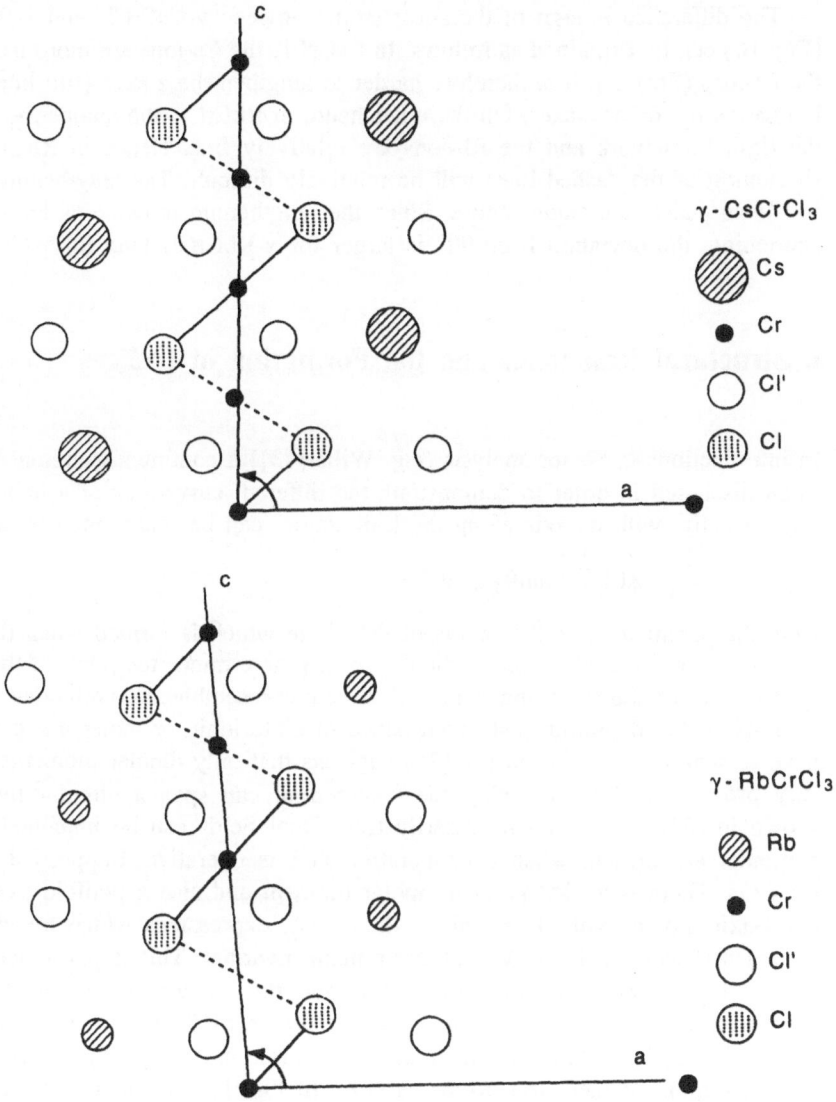

Fig. 16. The a, c-planes in γ-RbCrCl$_3$ and γ-CsCrCl$_3$. All ions lie in these planes, except the *open circles* which denote projections of Cl-ions

the orientation of the long axes with x and y. We define similar local axes in these octahedra, which are obtained from the first ones by reflection in the face sharing plane. The z-axis in these octahedra are indicated in the drawing by dashed lines. The orientations previously denoted by x and y can be expressed as $-1/2Q_\theta + 1/2\sqrt{3}Q_\varepsilon$ and $-1/2Q_\theta - 1/2\sqrt{3}Q_\varepsilon$.

Since both have $-1/2Q_\theta$ as components, the z-axes (dashed lines) will be shortened by half the amount by which the former z-axes (full lines) are lengthened.

The difference in sign of the shear strain between γ-CsCrCl$_3$ and γ-RbCrCl$_3$ (Fig. 16) can be explained as follows. In CsCrCl$_3$ the Cs-ions are more fixed than the Cl-ions (Sect. 3). It is therefore harder to lengthen the z-axes (full lines) than to shorten the other z-axes. On the other hand, in RbCrCl$_3$ the halogen ions form the rigid framework and the Rb-ions are relatively free. Hence in RbCrCl$_3$ the shortening of the dashed lines will be relatively difficult. The lengthening of the full lines makes the angle obtuse. Since the lengthening is twice as large as the shortening, the deviation from 90° is larger for γ-RbCrCl$_3$ than for γ-CsCrCl$_3$.

6 Structural Resonance and the Formation of Helices

In introductions to vector analysis (e.g. Wills [73]) a continuous circular helix is often discussed in order to demonstrate the different curvatures of a helical line. Such a helix, with its axis along the k-direction, can be represented by a vector

$$r = q\cos\theta\,\mathbf{i} + q\sin\theta\,\mathbf{j} + p\theta\,\mathbf{k} \ . \tag{25}$$

Here the parameter q is the radius of the circle which is formed when the helix is projected on to a plane perpendicular to $\mathbf{k}$. p determines the pitch of the helix.

Crystal structures having a helical structure resemble discontinuous helices, contrary to the definition just given, since discrete ionic or molecular groupings have to span the helix. Also Eq. (25) suggests that only dipolar moments, which vary properly in direction alng the k-direction, can span a circular helix. As shown in [53], this is not necessarily true. Dual fields can be imagined and in particular are present when a cooperative JTE is operative. Suppose $\theta = 0$ in Eq. (25). There is at that point a dipolar moment and also a twofold axis along the x-axis. For a twofold axis along the x-axis, expressions which transform as x^n or $(yz)^n$ where n is an odd integer remain invariant. This is particularly, true for the dipolar moment along the x-axis and the quadrupolar moment, which transforms as (yz). But choosing a different value of θ changes the directions. E.g., at $\theta = \pi/2$ the dipole moment and the twofold axis are along the y-axis and the quadrupole moment transforms as (xz). In Fig. 17 a circular helix has been sketched, with the dipole (single arrow) and the quadrupole moments (represented by a double arrow) at two general points. Note that the two types of arrows are perpendicular to each other.

From Sect. 2 we know that quadrupolar moments which transform as (xz) and (yz) are present in the hexagonal perovskites with copper. So one could imagine building a helix with only quadrupoles, provided these change properly in direction along the k-axis. This, however, is not realistic since there are no symmetry arguments which can prevent dipole moments being present as well. We have also shown in Sect. 2. that dipolar moments arise. These dipolar and quadrupolar moments constitute dual deformations, which go together in such a way that these are locally perpendicular to each other in the way just described.

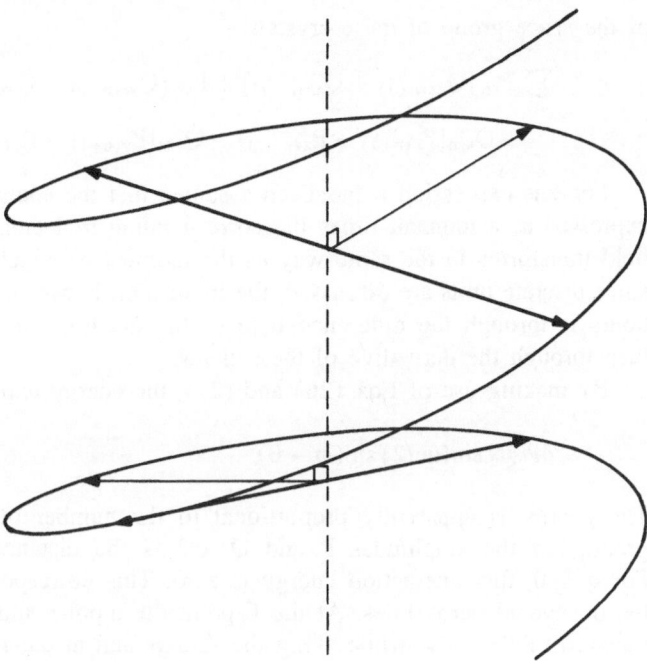

Fig. 17. Representation of a continuous helix with dipolar moments (*single arrows*) and quadrupolar moments (*double arrows*) at two arbitrary points

In this paper, we will restrict ourselves to one-threaded screws. Dual deformations of double helices and more general m-tuple helices have been described by Haije et al. [53] as well.

For a crystal with two face sharing octahedra per cell (the hexagonal perovskites) it is easy to describe the interaction between the dipolar and the quadrupolar waves, since harmonic theory can be used to illustrate the phenomenon of structural resonance for helices. Let the dipolar moments for a discontinuous helix be represented by

$$P_{xn} = P\cos(qz_n + \phi); \quad P_{yn} = -P\sin(qz_n + \phi). \tag{26}$$

Here n numbers the octahedra in a column along the c-direction. P is the amplitude of the wave and q the wave vector along the c^*-axis. ϕ is a phase angle which will be discussed later.

Let the quadrupoles with amplitude Q and phase θ be given by

$$Q_{xzn} = Q\cos(qz_n + \theta); \quad Q_{yzn} = -Q\sin(qz_n + \theta). \tag{27}$$

The interactions between these waves are proportional to the following expression, which can be shown to belong to the totally symmetric irreducible representation of the space group of these crystals:

$$\sum_n P_{xn} [Q_{xz(n+1)} - Q_{xz(n-1)}] + P_{yn} \{Q_{yz(n+1)} - Q_{yz(n-1)}]$$

$$-Q_{xzn}[P_{x(n+1)} - P_{x(n-1)}] - Q_{yzn}[P_{y(n+1)} - P_{y(n-1)}]. \qquad (28)$$

For this expression it has been assumed that the energy of each type can be expressed as a moment times the corresponding molecular field. This molecular field transforms in the same way as the moment to which it belongs. Note that, since discrete units are discussed, the interaction between, say, P_{xn} and its neighbours is through the difference field of the quadrupolar molecular fields rather then through the derivative of these fields.

By making use of Eqs. (26) and (27), the energy expression becomes

$$4PQN \sin(qc/2) \sin(\phi - \theta). \qquad (29)$$

The energy is apparently proportional to the number of sites (N) and to the product of the amplitudes P and Q. c/2 is the distance between neighbours. For $q = 0$, this interaction energy is zero. This corresponds to the situation in the hexagonal perovskites. At the Γ-point the dipolar and quadrupolar moments transform differently, while along the Δ-axis and at the A-point the dipolar and quadrupolar moments transform with the same irreducible representation.

The absolute value of Eq. (29) is maximal for $q = \pi/c$ (at the A-point) and for the difference in phases $(\phi - \theta) = \pm\pi/2$. Apparently there are two solutions, which have the dipole and the quadrupolar wave 90° out of phase, the lower one leading to stabilization. This is the phenomenon of 'Structural Resonance', first described by Heine and McConnell [24, 25] for incommensurate structures and by Maaskant and Haije [26] for CsCuCl$_3$, Haije and Maaskant [27] for TMCuCl$_3$, and for helices in Se/Te, HgO/HgS/HgSe by Haije et al. [53].

The fact that the dipole and the quadrupolar mode are 90° out of phase (see Fig. 17), can also be shown by drawing projections of the important octahedra on a plane perpendicular to the axis of the helix. In all cases which we considered, these octahedra would have inversion symmetry when not distorted. The observed deformations can then be decomposed into even and odd parts with respect to inversion. In all cases these are perpendicular to each other in the sense discussed as for Fig. 17. We refer for these drawings to the original papers.

The reason for the interaction between dipole waves and quadrupole waves of the right symmetry can also be illustrated by drawing at distances c/2 p_x and d_{xz}-orbitals (Fig. 18). It is evident that for waves with wavelength 2c overlap is not zero and therefore interaction is expected.

The stabilization of a helix with respect to deformations with a single direction, e.g. only P_x and Q_{zx}, can be shown to depend on fourth order anharmonic terms. A helix has less tension than a planar deformation with the same harmonic energy [26]. In the case of the cooperative JTE local third order anharmonic terms, which for CsCuCl$_3$ result in a third order Landau invariant of the crystal distortion mode, these also favour a helical distortion. Through this term the

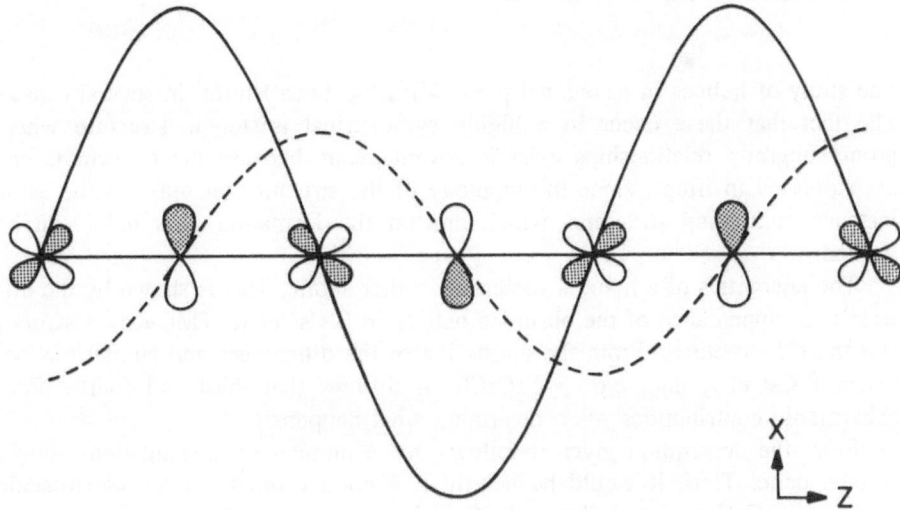

Fig. 18. Principle of the stability of a helical structure. Dipoles along the x-direction (denoted as p_x-functions) and quadrupoles, transforming as d_{xz}-functions, are arranged in an antiferrodistortive order along the z-axis. Their molecular fields are denoted as sine curves. The quadrupoles and dipoles can interact by orienting in the gradient of the molecular field of the other. In circular helix there is a similar arrangement of moments in the yz-plane, but shifted over $\pm\ 90°$ with respect to the xz-plane

Table 2. Synopsis of the studied helices.

Compound	Parent group	B.Z	Group of k	Irrep	Dim	Ref.
CsCuCl$_3$	P6$_3$/mmc	Δ	C$_{6v}$	E$_1$	2	16.26
TMCuCl$_3$	P6$_3$/mmc	Δ	C$_{6v}$	E$_1$, E$_2$	2	27
CsCuBr$_3$	P6$_3$/mmc	A	D$_{6h}$	F	4	–
C2-structure	P6$_3$/mmc	A	D$_{6h}$	F	4	–
Se/Te	Pm3m	Λ	C$_{3v}$	E	2	53
HgO; HgS; HgSe	Fm3m	Λ	C$_{3v}$	E	2	53

distortion wave is pinned to the lattice, which means that the remaining phase angle in Eq. (29) is fixed.

In Table 2, helical compounds with parallel axes, which can be shown to be derivable by one or two 'soft modes' from a crystal structure of high symmetry, are listed together with the information of the irreps (irreducible representations). These modes share the property of belonging to two-fold degenerate irreps. This is understandable since one wants to let dipole moments in the x- and y-direction (the axis of the helix is along the z-direction) to behave similarly. In all given cases the irreps of the little group is of the E-type, except for CsCuBr$_3$ and the C2 structures, where it is four-dimensional. Essential is whether two directions (e.g. x and y) are degenerate, which – as is seen in a system with non-parallel axes – does not mean that the little group should have degeneracy.

7 Discussion and Conclusions

The study of helices in hexagonal perovskites has been fruitful in several senses. The fact that these occur in a highly symmetrical aristotype structure where group-subgroup relationships exist is advantageous because the distortions can be expressed in irreps, since the topology of the structure remains. At the same instance competing structures which prevent the formation of a helix can be studied.

The energetics of a helix is sometimes rather subtle. This is shown by the difference in appearance of the observed helices in β-CsCuCl$_3$, TMCuCl$_3$, CsCuBr$_3$ and the C2-structures. From the discussion of the differences and similarities between β-CsCuCl$_3$ and, e.g., γ-CsCrCl$_3$, it follows that third and fourth order anharmonic contributions often determine what happens.

From the description given it follows that a number of investigations should still be done. Thus, it would be helpful if a good estimate of the electrostatic energy of β-CsCuCl$_3$ could be made. Cr^{2+} has been studied by acoustic paramagnetic resonance in several oxides, but never in chlorides.

It seems unlikely that a general physical theory can explain all phenomena in such a group of crystals with the same aristotype structure. There are many surprising effects. In Sect. 2 we emphasized the easy directions of the shifts of close packed layers. Because of the ion sizes we would have expected CsCuBr$_3$ to be similar to RbCuCl$_3$. But dimerization and a Spin-Peierls effect occurs in CsCuBr$_3$. The difference between CsCuCl$_3$ and the Cr-compounds has to be ascribed to the difference in magnetic properties of Cr^{2+} and Cu^{2+}. Also the structure of TMCuCl$_3$ exhibits a delicate balance between free energy contributions, with the temperature playing an important role.

Acknowledgements. The author wishes to thank Dr. D.J.W. IJdo and Dr. R.A.G. de Graaff for their continuing help in preparative solid state chemistry and in the interpretation of diffraction patterns of the crystals, respectively.

8 Appendix

In this appendix the transformation is given for the ε_g-normal modes of a regular octahedron (O$_h$) in the usual representation x, y, z (small letters, which denote displacements of ions from their equilibrium positions) into a X, Y, Z cartesian axes system with the Z-axis along the [111] and the Y-axis along [1, -1, 0]. The numbering of the six ligands is conventional (Sugano et al. [75]). In the original axes system the expression for the ε_θ and the ε_ε normal modes is given by

$$Q_\theta = 1/\sqrt{12}\{2z_3 - 2z_6 - x_1 + x_4 - y_2 + y_5\} \tag{A1}$$

$$Q_\varepsilon = 1/2\{x_1 - x_4 - y_2 + y_5\} . \tag{A2}$$

The transformation between the xyz and the XYZ axes systems is given by

$$x = -X/\sqrt{6} + Y/\sqrt{2} + Z/\sqrt{3}$$
$$y = -X/\sqrt{6} - Y/\sqrt{2} + Z/\sqrt{3}$$
$$z = X\sqrt{2/3} + Z/\sqrt{3} . \tag{A3}$$

The deformations of the octahedron are described in terms of the triangles A and B, constituted respectively by the ligands 1, 2, 3 and the ligands 4, 5, 6. The changes of the triangles are shifts (S), tilts (T) and deformations (D). These have the following normalized expressions:

$$S_{Ax} = (X_1 + X_2 + X_3)/\sqrt{3}; \quad S_{Ay} = (Y_1 + Y_2 + Y_3)/\sqrt{3} \tag{A4}$$

$$T_{Ax} = (Z_1 - Z_2)/\sqrt{2}; \quad T_{Ay} = (-2Z_3 + Z_1 + Z_2)/\sqrt{6} \tag{A5}$$

$$D_{Ax} = (-\tfrac{1}{2}X_1 - \tfrac{1}{2}\sqrt{3}Y_1 - \tfrac{1}{2}X_2 + \tfrac{1}{2}\sqrt{3}Y_2 + X_3)/\sqrt{3} \tag{A6}$$

$$D_{Ay} = (\tfrac{1}{2}\sqrt{3}X_1 - \tfrac{1}{2}Y_1 - \tfrac{1}{2}\sqrt{3}X_2 - \tfrac{1}{2}Y_2 + Y_3)/\sqrt{3} \tag{A7}$$

$$S_{Bx} = (X_4 + X_5 + X_6)/\sqrt{3}; \quad S_{By} = (Y_4 + Y_5 + Y_6)/\sqrt{3} \tag{A8}$$

$$T_{Bx} = (Z_4 - Z_5)/\sqrt{3}; \quad T_{By} = (2Z_6 - Z_4 - Z_5)/\sqrt{6} \tag{A9}$$

$$D_{Bx} = (+\tfrac{1}{2}X_4 + \tfrac{1}{2}\sqrt{3}Y_4 + \tfrac{1}{2}X_5 - \tfrac{1}{2}\sqrt{3}Y_5 - X_6)/\sqrt{3} \tag{A10}$$

$$D_{By} = (-\tfrac{1}{2}\sqrt{3}X_4 + \tfrac{1}{2}Y_4 + \tfrac{1}{2}\sqrt{3}X_5 + \tfrac{1}{2}Y_5 - Y_6)/\sqrt{3} \tag{A11}$$

$$Q_\theta = (S_{Ax} - S_{Bx} + D_{Ax} + D_{Bx} - T_{Ay} - T_{By})/\sqrt{6} \tag{A12}$$

$$Q_\varepsilon = (S_{Ay} - S_{By} - D_{Ay} - D_{By} + T_{Ax} + T_{Bx})/\sqrt{6} . \tag{A13}$$

9 References

1. Schlueter AW, Jacobson RA, Rundle RE (1966) Inorg Chem 5: 277–280
2. Weenk JW, Spek AL (1976) Cryst Struct Comm 5: 805–810
3. Willet RD, Bond MR, Haije WG, Soonieus OPM, Maaskant WJA (1989) Inorg Chem 27: 614–620 (see also 28: 810)
4. Ting-i Li, Stucky GD (1973) Inorg Chem 12: 441–445
5. Kroese CJ, Tindemans van Eyndhoven JCM, Maaskant WJA (1971) Solid State Commun 9: 1707–1709
6. Kroese CJ, Maaskant WJA, Verschoor GC (1974) Acta Crystallogr B30: 1053–1056
7. Crama WJ (1981) Acta Crystallogr. B37: 2133–2136
8. Crama WJ (1980) Thesis University of Leiden; The Netherlands
9. Lüthi B (1969) Private communication (see [16])
10. Graf HA, Tanake H, Dachs H, Pyka N, Schotte U, Shirane G (1986) Solid State Commun 57: 469–472
11. Graf HA, Shirane G, Schotte U, Dachs H, Pyka N, Iizumi M (1989) J Phys; Condens. Matter 1: 3743–3763
12. Schotte U (1987) Z Phys B66: 91–101
13. Schotte U, Graf HA, Dachs H (1989) J Phys; Condens Matter 1: 3765–3787
14. Pyka N, Schotte U, Graf HA, Dachs H (1989) Solid State Commun 70: 1105–1108

15. Hirotsu S (1975) J Phys C: Solid State Phys 8 L: 12–16
16. Kroese CJ, Maaskant WJA (1974) Chem Phys 5: 224–233
17. Hirotsu S (1977) J Phys C: Solid State Phys 10: 967–985
18. Hirakawa K, Yamada I, Kurogi Y (1970) Proc Int Conf Magn, Grenoble (J Phys 32, Suppl. C-1: 890–891 (1971)
19. Reinen D, Friebel C (1979) Structure and Bonding, Springer Verlag, Berlin, Vol. 37, pp 1–60
20. Tindemans-van Eyndhoven JCM, Kroese CJ (1975) J Phys C: Solid State Phys 8: 3963–3974
21. Kroese CJ (1976) Thesis University of Leiden; The Netherlands
22. Tindemans-van Eyndhoven JCM (1978) Thesis University of Leiden; The Netherlands
23. Petitgrand D, Hennison B, Escribe-Filippini C, Legrand S (1981) J Phys C-6, 42: 782–784
24. Heine V, McConnell JDC (1984) J. Phys. C: Solid State Phys. 17: 1199–1220
25. McConnell JDC, Heine V (1984) Acta Crystallogr. A40: 473–482
26. Maaskant WJA, Haije WG (1986) J Phys C: Solid State Phys 19: 5259–5308
27. Haije WG, Maaskant WJA (1988) J Phys C: Solid State Phys 21: 5337–5350
28. Laiho R, Natarajan M, Kaira M (1973) Phys. Status. Solidi a15: 311–317
29. Fernández J, Tello MJ, Peraza J, Bocanegra EH (1976) Mat Res-Bull 11: 1161–1168
30. Khomskii DL (1977) JETP Lett. 25: 544–546
31. Lee BS (1979) J Phys C: Solid State Phys 12: 855–863
32. Vasudevan S., Shaikh AM, Rao CNR (1979) Phys Lett 70A: 44–46
33. Tanaka H, Dachs H, Iio K, Nagata K (1986) J Phys C: Solid State Phys 19: 4861–4878
34. Tanaka H, Dachs H, Iio K, Nagata K (1986) J Phys C: Solid State Phys 19: 4879–4896
35. Tanaka H, Iio K, Nagata K (1981) J Phys Soc Japan 50: 727–728
36. Tanaka H, Tazuke Y, Iio K, Nagata K (1983) J Phys Soc Japan 52: Suppl. 76
37. Tazuke Y, Tanaka H, Iio K, Nagata K (1981) J Phys Soc Japan 50: 3919–3924
38. Tazuke Y, Tanaka H, Iio K, Nagata K (1984) J Phys Soc Japan 53: 3991–3197
39. Gesi K, Osawa K (1984) J Phys Soc Japan 53: 907–909
40. Crama WJ, Maaskant WJA (1983) Physica 121 B+C: 219–232
41. Höck KH, Schröder G, Thomas H (1978) Z Phys B30: 403–413
42. Crama WJ (1981) J Solid State Chem 39: 168–172
43. Harada M (1982) J Phys Soc Japan 51: 2053–2054
44. Harada M (1983) J Phys Soc Japan 52: 1646–1657
45. Harada M, Fischer JE, Shirane G, Yamada Y (1987) J Phys Soc Japan 56: 3786–3788
46. Crama WJ, Maaskant WJA, Verschoor GC (1974) Acta Crystallogr B34: 1973–1974
47. Haije WJ, Frikkee E, Doorenbos G, Maaskant WJA, Visser D (1990) "Netherlands Energy Research Foundation E.C.N.". Quaterly Progress Report of the Physics Department, period 1-9-'88 to 31-12-'88, compiled by J Bergsma
48. Harrison A, Visser D (1989) J Phys: Cond Matter 1: 733–754
49. Jouini N, Guen L, Tournoux M (1982) Mat Res Bull 17: 1421–1427
50. Visser D, Verschoor GC, IJdo DJW (1980) Acta Crystallogr B36: 28–34
51. Pauling L (1960) Nature of the Chemical Bond, 3rd ed., Cornell University Press, Ithaca New York
52. Shannon RD, Prewitt CT (1970) Acta Crystallogr. B25: 925–946 (1969) and B26: 1046–1048
53. Haije WG, Dobbelaar JAL, Welter MH, Maaskant WJA (1987) Chem Phys 116: 159–170 (1987)
54. Stucky GD, D'Agostino S, McPherson G (1966) J Am Chem Soc 88: 4828–4831
55. McPherson GL, McPherson AM, Atwood JL (1980) J Phys Chem Solids 41: 495–499
56. McPherson GL, Stucky GD (1972) J Chem Phys 57: 3780–3786
57. Ting-i Li, Stucky GD (1973) Acta Crystallogr B29: 1529–1532
58. Goodyear J, Kennedy DJ (1972) Acta Crystallogr B28: 1640–1641
59. Seifert HJ, Dau E (1972) Z Anorg Allg Chemie 391: 302–312
60. Cox DE, Merkert FC (1972) J Cryst Growth 13/14: 282–284
61. Crama WJ, Bakker M, Verschoor GC, Maaskant WJA (1979) Acta Crystallogr B35: 1875–1877
62. Leech DH, Machin DJ (1974) J.C.S. Chem Com 866–867
63. Leech DH, Machin DJ (1975) J.C.S. Dalton 1609–1614
64. Larkworthy LF, Trigg JK (1970) J.C.S. Chem Com 1221–1222
65. Perez-Mato JM, Tello MJ, Manes JL, Bocanegra EH, Fernandez J (1980) Ferroelectrics 25: 457–459
66. Perez-Mato JM, Manes JL, Tello MJ (1980) J Phys C: Solid State 13: 2667–2674
67. Kahn O (1985) Z. Angewandte Chemie 24: 834–850
68. Niel M, Cros C, Pouchard M, Chaminade JP (1977) J Solid State Chem 20: 1–8

69. McPherson GL, Sindel LJ, Quarls HF, Frederick CB, Doumit CJ (1975) Inorg Chem 14: 1831–1834
70. Abragam A, Bleaney B (1970) Electron Paramagnetic Resonance of Transition Ions, Clarendon Press, Oxford
71. Tazuke Y, Kinouchi S, Tanaka H, Iio K, Nagata K (1986) J Phys Soc Japan 55: 4020–4024
72. Bersuker IB, Polinger VZ (1989) Vibronic Interactions in Molecules and Crystals, Springer Verlag, Berlin
73. Wills AP (1958) Vector Analysis with an Introduction to Tensor Analysis, Dover Publications, Inc., New York
74. Megaw HD (1973) Crystal Structures: A Working Approach, W.B. Saunders Company, Philadelphia
75. Sugano S, Tanabe Y, Kamimura H (1970) Multiplets of Transition Metal Ions in Crystals, Academic Press, New York Fig. 6.2
76. Crama WJ, Zandbergen HW (1981) Acta Crystallogr. B37: 1027–1031
77. Zandbergen HW, IJdo DJW (1980) J Solid State Chem 34: 65–70
78. Zandbergen HW, IJdo DJW (1981) J Solid State Chem 38: 199–210
79. Garcia J, Bartelome J, Navarro R, Gonzales D, Crama WJ, Maaskant WJA (1981) In: Devreese JT Recent Dev Condens Matter Phys [Pap Gen Conf] 1st 1980 Plenum New York, N.Y., Vol.4, pp.11–20.
80. Schläfer HL, Gliemann G (1967) Einführung in die Ligandenfeld Theorie: Akademische Verlagsgesellschaft, Frankfurt am Main
81. Kauzmann W (1957) Quantum Chemistry, Academic Press, New York
82. e.g. Fieser LF, Fieser M (1956) Organic Chemistry, Reinhold, New York
83. Ham F.S (1972) In: Geschwind S (ed) Electron Paramagnetic Resonance, Plenum Press New York
84. Ham F.S (1987) Phys Rev Lett 58: 725–728
85. Alcock NW, Putnik CF, Holt SL (1976) Inorg Chem 15: 3175–3178
86. Adachi K, Achiwa N, Mekata M (1980) J Phys Soc Japan 49: 545–553

Author Index Volumes 1-83

Springer-Verlag
and the Environment

We at Springer-Verlag firmly believe that an international science publisher has a special obligation to the environment, and our corporate policies consistently reflect this conviction.

We also expect our business partners – paper mills, printers, packaging manufacturers, etc. – to commit themselves to using environmentally friendly materials and production processes.

The paper in this book is made from low- or no-chlorine pulp and is acid free, in conformance with international standards for paper permanency.